Environmental Ethics

ALSO AVAILABLE FROM BLOOMSBURY

Ethics of Climate Change, James Garvey

Eco-Aesthetics, Malcolm Miles

Religion in Environmental and Climate Change, Dieter Gerten

Forthcoming titles

Climate Ethics, Sarah Kenehan

Environmental Ethics

From theory to practice

MARION HOURDEQUIN

BLOOMSBURY ACADEMIC
LONDON • NEW YORK • OXFORD • NEW DELHI • SYDNEY

BLOOMSBURY ACADEMIC
Bloomsbury Publishing Plc
50 Bedford Square, London, WC1B 3DP, UK
1385 Broadway, New York, NY 10018, USA
29 Earlsfort Terrace, Dublin 2, Ireland

BLOOMSBURY, BLOOMSBURY ACADEMIC and the Diana logo
are trademarks of Bloomsbury Publishing Plc

First published 2015
Reprinted by Bloomsbury Academic 2016 (four times), 2017 (twice), 2018 (three times),
2019, 2021, 2022, 2023

For legal purposes the Acknowledgements on p. x constitute
an extension of this copyright page.

A catalogue record for this book is available from the British Library.

ISBN: HB: 978-1-47250-808-9
PB: 978-1-47251-098-3
ePDF: 978-1-47250-783-9
ePub: 978-1-47250-761-7

Hourdequin, Marion.
Environmental ethics : from theory to practice / Marion Hourdequin.
pages cm
ISBN 978-1-4725-0808-9 (hardback) – ISBN 978-1-4725-1098-3 (paperback)
1. Environmental ethics. I. Title.
GE42.H68 2015
179'.1–dc23
2014031371

Typeset by Newgen Knowledge Works (P) Ltd., Chennai, India
Printed and bound in Great Britain

To find out more about our authors and books visit
www.bloomsbury.com and sign up for our newsletters.

For David, Adele, and Timothy, in gratitude
And to my parents, for the opportunity to learn

Contents

List of Illustrations

Figures

Table

Acknowledgments

A number of people played important roles in the development and refinement of this book. I am particularly grateful to David Havlick, Rick Furtak, and Julia Liao. David Havlick read the manuscript in its entirety, providing helpful feedback on all of the chapters. Julia Liao provided comments on the book and developed almost all of the material for the book's companion website. My colleague Rick Furtak generously read and helpfully commented on many chapters. In addition, I am grateful to my environmental ethics classes, and to Alan Holland, Katie McShane, and two anonymous reviewers for suggestions and advice. I also thank the editors at Bloomsbury and production team at Newgen Knowledge Works for their help throughout the project. Finally, my family showed tremendous patience and support in seeing this book to completion, and for that and much else, I am grateful.

PART ONE

Environmental values

1

Bringing values to light

Introduction

We live on a beautiful and finite planet, and through our lives, we are changing the planet in many ways. Change is inevitable, with or without us. The history of our planet is a history of change, and over geologic time, a history of radical change. Yet with every change, there is loss. Some losses are relatively insignificant: when I set down a pencil on a page and begin to draw a picture, the blank page is lost—but ideally, transformed into something new and interesting. The prospect of advancement motivates many of the changes we make to our environment: we cut forests to build houses and mine metals to build newer and more powerful computers and phones. Some of these changes generate goods, but also involve significant losses. Some of these losses are unintended or unknown as we pursue change and improvement. When we first began to burn fossil fuels, for example, we knew nothing of global climate change, though this now has become one of the most serious and daunting environmental problems we face.

Environmental ethics is not about halting progress or stopping change. It is, however, about reflecting on what constitutes progress, the kinds of changes we should pursue, and how we should respond to the inevitable losses we produce through our actions. Environmental ethics—though it questions the status quo and subjects our actions to critical evaluation—is a fundamentally hopeful discipline. It is hopeful because it rests on the conviction that we can do better. As reflective, thinking beings, we can learn from our successes and our failures, and as empathic, caring beings, we can take into account the well being of other persons, other living things, and our planet as a whole.

One of the hopeful strands of environmental ethics derives from the ancient Greek tradition of virtue ethics, in which Aristotle's philosophy looms large. Although some of his views are not widely shared today, one of Aristotle's fundamental convictions remains deeply salient. It is the conviction that a good life for human beings is compatible with, and indeed is constituted by, a life of virtue. In his *Nicomachean Ethics*, Aristotle explains how human flourishing depends on the fulfillment of our potential as rational and social beings. Contemporary environmental virtue ethicists suggest that in today's world, human flourishing depends similarly on the development of our full potential as reflective and compassionate beings, and particularly on our ability to live well not only with respect to other human beings, as members of the social world, but also with respect to other living things, as members of the natural world. It is one of the aims of this book to encourage reflection on this possibility and consider how it might be realized. If we can envision a way in which we can live happy and fulfilling lives while respecting the broader world around us, that will be a key step in moving toward sustainable economies and livelihoods, and to tackling the many environmental challenges we face in the decades to come.

In tackling environmental problems, however, many people assume that we need only science, or economics, or both, in order to find solutions. It is a second important aim of this book to dispel this assumption and show the critical place of philosophical reflection—and ethical reflection in particular—in addressing the challenges ahead. The idea that sound science and rigorous economic analysis are sufficient to resolve environmental controversies is too simple. As Daniel Sarewitz (2004, p. 391) points out regarding controversies over genetically modified organisms (GMOs), scientists from different disciplines tend to focus on different aspects of GMOs:

> It . . . seems reasonable to expect that scientists from disciplines involved in design and application of GMOs, such as plant geneticists and molecular biologists, would be potentially more inclined to view GMOs in terms of their planned benefits, and ecologists or population biologists would be more sensitized to the possibility of unplanned risks at a systemic level.

Sarewitz's comment highlights the way in which different scientific disciplines operate using different frameworks and background assumptions, and reflect different values and priorities. Similarly, economic analysis is not value neutral (see DesJardins 2006, ch. 1, for further discussion). We find this illustrated in the fact that economic assessments of the impacts of global climate change and the benefits of acting to forestall it have reached a wide array of conclusions, depending on the assumptions and values driving the analyses.

Although there are many ways to argue for the importance of moral philosophy to contemporary life, the chief reason to be emphasized here is that values are ubiquitous in science, economics, and policy, including environmental science, environmental economics, and environmental policy.

Philosophy is important in addressing environmental problems because philosophy can help bring these values to light, where they can be better discussed and evaluated. When values remain hidden, they guide our decisions without our awareness that they are doing so. The next section illustrates the ubiquity of values, paying particular attention to the role of values in science and economics, and the relevance of these values to environmental issues.

The ubiquity of values

Values in science

If you took science classes in secondary school, you most likely studied the scientific method. In studying this method, you probably learned that the best kind of science is objective and value free. A hallmark of bad science is bias, and the scientific method seeks to eliminate it. The scientific method, ideally, enables any investigator—irrespective of her political persuasion, religious views, cultural background, philosophical outlook, or value commitments—to study a problem, conduct an experiment, and generate the same (or very similar) results as any other investigator.

Most philosophers would agree that science is a systematic mode of investigation that aims to make its methods as transparent and replicable as possible, and that aims to minimize the influence of extraneous factors and biases on its results. Nevertheless, in the wake of convincing arguments by Thomas Kuhn, author of *The Structure of Scientific Revolutions*, many philosophers and science studies scholars have come to doubt that science can ever be truly and fully value free. Kuhn argued that science operates within paradigms. A *scientific paradigm* is a mode of scientific research organized around particular key questions, methodological approaches, assumptions, and evaluative criteria. Ptolemaic astronomy, for example, operated under the basic assumption that the earth was at the center of the universe, and this assumption shaped inquiry by generating particular kinds of questions: "[Given that the earth is at the center of the universe,] how can we explain the orbit of Mars?" The bracketed text here indicates the basic background assumption, under which the more specific question about Mars could be asked and answered.

Kuhn argued that science *needs* paradigms. As he explained, "In the absence of a paradigm . . . all of the facts that could possibly pertain to the development of a given science are likely to seem equally relevant" (Kuhn 1996, p. 15). Kuhn's argument was that scientific investigation always takes place in the context of certain key background assumptions about how the world works. To elucidate with a simple example, imagine that my tomato plants are not flourishing: their leaves are wilted and yellow, and I want to know why. In developing hypotheses about the problem with my tomato plants, I draw on a whole host of background assumptions. I assume that there are certain critical elements essential for plants to thrive: light, nutrients, and water. I assume that a deficiency of any of these elements can reduce growth and health. In addition, I believe that insects, fungi, and herbivorous animals such as squirrels and deer sometimes damage plants. Those assumptions guide my thinking about what might be wrong with my plants. I rule out—without even considering it—the hypothesis that the noise of my lawnmower or the sound of my children's voices in the yard is bothering the plants. Similarly, I do not bother to explore the possibility that the vibrations from cars driving down the alley behind my house are damaging my plants. I generate my guesses about what is wrong with my tomato plants in light of a background theory regarding the key elements that facilitate or undermine the health of plants.

We can see, then, that certain background assumptions influence the conduct of science and the trajectory of scientific investigations. Scientists use these background assumptions to identify those hypotheses that are most plausible, and it is these that they typically pursue. But what does all of this have to do with values? We can return again to the idea of paradigms to help answer this question, for paradigms are inevitably value laden. For example, Kuhn emphasized that a paradigm includes certain methods of investigation such as the use of particular instruments and techniques. If we look to the sixteenth century, we see scientists such as René Descartes arguing that animals are similar to machines and lack a soul. This view, in turn, was clearly fueled by a broader understanding of animals' place in the natural order and their distinctness from human beings in this regard. These ideas reflected a scheme of valuation in which humans have special and significant moral value as compared to the lesser value of animals. In keeping with these ideas, vivisection—the live dissection of animals—was common in Descartes' time. Today, although inhumane treatment of animals remains widespread, there are significantly greater restrictions on animal experimentation. These restrictions shape the kinds of research that are conducted and the kinds of questions that can be easily asked and answered.

In reflecting on these attitudinal changes in animal research, it is worth noting that science in this case did not change from value laden to value free. Instead, the values that guided science changed: we no longer think it justified, in general, to experiment on live, unanaesthetized animals. What is important, one might argue, is not that values be purged from science entirely, but that they be subject to critical scrutiny. Contemporary philosophers of science, such as Helen Longino (1990) and Philip Kitcher (2001), argue that science depends on values to function well, yet they also aim to make clear the roles values play in science, so that we can reflectively consider which values science ought to embody. Longino (1990) has pointed out that science operates according to certain constitutive values, while also being influenced by contextual values. Broadly speaking, Longino is distinguishing between the internal values of scientific practice (constitutive values) and the external, social values that characterize the broader context in which science takes place (contextual values). Constitutive values include epistemic (knowledge-related) values such as accuracy, consistency, and coherence, while contextual values might include factors such as social justice or an emphasis on economic growth.

However, the distinction between contextual and constitutive values is itself hard to maintain, because the two are clearly related in important ways. For example, in keeping with the constitutive values of scientific practice, scientists typically seek very strong evidence before rejecting a null hypothesis, or default assumption about the relationship between two variables or the nature of some phenomenon. However, as we shall see, these constitutive values do not always support certain objectives, such as the preservation of endangered species. Constitutive values are not neutral in relation to social values, and in certain cases, it may be appropriate to adjust scientific practice and modify constitutive values in order to better respond to contextual values.

To take a more specific case, in population ecology, scientists study trends in population sizes of various organisms and attempt to understand the basis for these trends. A population ecologist studying lynx in the northern Rocky Mountains of the United States might seek to know whether lynx are declining, increasing, or stable in number. A typical null hypothesis would hold that lynx populations are stable, and significant data would then be required to dislodge this hypothesis and support the conclusion that lynx are in decline. In fact, using widely accepted scientific standards, the scientist would need to show that the probability of his or her observational data given a stable, nondeclining lynx population was less than 5 percent. In other words, the scientist can reject the null hypothesis and accept the alternative hypothesis (that lynx are in decline) only if there is a 95 percent chance that the data are indicative of a declining population.

Rejecting the null hypothesis thus requires clearing a high bar, and the justification for this high bar rests with the structure of science. Science is a cumulative enterprise, and although science progresses in some cases by overturning prior theories, it also builds on prior conclusions, and the reliability of these conclusions is highly valued. Notice, though, that in the case of populations declining toward extinction, the standard, conservative approach to rejecting the null hypothesis may make it difficult to justify species-saving intervention before it is too late. That is, by the time we have sufficient data to conclude that a particular population is in decline, the decline may be too serious to reverse. This has led philosophers as well as biologists to argue that standards of proof need to be context-sensitive (see Shrader-Frechette and McCoy [1993] and Taylor and Gerrodette [1993] for further discussion). Where the costs of a particular kind of error are high, standards should be adjusted to account for these costs. In the case of endangered species, it may sometimes be better to risk erroneously concluding that a population is in decline than to erroneously assume that it is stable. It would be better to avoid error altogether, of course, but in a world of limited information and limited time, value judgments are required to determine how best to treat uncertainty in science, and how best to act on the limited knowledge available.

If all this sounds complicated, that is because it is. Although we may easily agree that certain kinds of blatant bias in science are unacceptable, the question of which values should guide science is much more complex, and thus ripe for philosophical analysis and reflection. This task is particularly relevant to environmental philosophy, because environmental policy decisions typically rely heavily on the natural sciences. We depend on population ecology to detect threatened and endangered species, on chemistry, toxicology, and health sciences to determine unhealthy levels of air pollution, and on atmospheric chemistry, physics, geology, biology, and many other disciplines to discern the likely effects of global climate change. All of these sciences rely on key methodological assumptions and on particular judgments regarding evidentiary standards and the treatment of uncertainty. Moreover, the task of bringing values to light and subjecting them to critical reflection does not end with science. Virtually every academic discipline, decision procedure, and political outlook reflects certain values, many of which go unstated. Environmental philosophers have been particularly concerned not only with values in science, but also with the way certain value assumptions undergird economics, which in turn guides many policy decisions.

Values in economics

Just as values inevitably guide science, values play an important role in economics. However, although these values may have significant consequences for environmental decision-making, they are not necessarily explicitly identified or discussed. Let us take a simple example to illustrate how value assumptions can be embedded in economic arguments, though not explicitly stated:

P1. The economic costs of saving the black-footed ferret exceed the economic benefits of doing so.

C1. We should not save the black-footed ferret.

Chapter 2 discusses cost–benefit analysis and other aspects of economic valuation in more depth, but for now, we can make a few important observations. It is also appropriate at this point to introduce some basic philosophical terminology in relation to arguments. An argument, philosophically speaking, provides a reasoned defense of some claim. Typically, an argument involves a logical chain of reasoning from premises, or assumptions, to a conclusion. The above-stated black-footed ferret argument is a very simple one. It contains a single premise (P1) and a single conclusion (C1). In assessing this argument, one might ask whether the conclusion logically follows from the premise. If the economic costs of saving the black-footed ferret exceed the economic benefits of doing so, does it logically follow that we should not save the black-footed ferret? Not exactly. In order for the conclusion to follow, we need to add a second premise, such as P2:

> P2. If the economic costs of taking a particular action exceed its economic benefits, one ought not take that action.

With this additional premise (known as a "suppressed" or implied, but unstated premise) the conclusion now follows logically. If P1 and P2 are true, then the conclusion C1 must also be true. An argument of this kind, where the conclusion follows directly and logically from the premises, is known as a valid, deductive argument. Validity refers to the logical form of an argument: an argument that is valid has a proper logical form, such that the truth of the premises guarantees the truth of the conclusion.

But even with an additional premise that establishes its logical validity, the ferret argument may be problematic. For while the conclusion follows from the premises, one or more of the premises might be false. Let us focus on the second premise, P2. This originally unstated premise is the one that most obviously involves value assumptions. Why, one might ask, ought we never to do something whose economic costs exceed its economic benefits? Perhaps, there are *non*economic values that weigh in favor of saving the ferret, even if the costs outweigh the benefits in strictly economic terms. For example, we might have a special obligation to try to save the ferret by virtue of the fact that human actions—associated with European settlement of the North American plains and the removal of prairie dogs, the ferret's chief prey—played a significant role in the ferret's decline. Or, maybe the ferret's aesthetic value needs to be counted more thoroughly than the economic calculation allows.

This discussion raises another interesting and challenging issue. In response to these "noneconomic" reasons, a defender of P2 might argue that economics *can* take account of such reasons, because all relevant costs and benefits can, in principle, be expressed in economic terms. A defender of this approach might acknowledge that capturing all the relevant costs and benefits requires special effort: merely looking at the way the ferret figures into existing economies might not fully capture its value. However, we could try to capture other values in economic terms through a variety of methods, which we will discuss in more detail in Chapter 2. For example, we might ask people how much they would be willing to pay to see a black-footed ferret in the wild, even if they never actually intend to do so, and we might develop further strategies to translate into economic terms the costs of failing to meet our moral obligations. Nevertheless, the claim that all costs and benefits can be adequately captured in economic terms is controversial. So even if one agrees that one should always choose an action whose benefits exceed its costs, one might object to P2, which focuses on economic benefits and costs. Relatedly, one might resist the idea that decisions should be determined by summing up the costs and benefits. We do not need to settle these questions here, but they help establish an important point. P2 is controversial, and it is therefore worthy of philosophical discussion and debate. This reiterates the point that by bringing values to light, they can be better evaluated and discussed.

For further thought

1. According to Thomas Kuhn, how do paradigms shape the conduct of science? Describe an example (real or fictional) in which a paradigm encodes certain values.
2. If values are inevitably part of science, how can we distinguish legitimate values from illegitimate ones?
3. What kinds of values is an economic cost–benefit analysis most likely to capture? What kinds of values might it be most likely to miss?
4. Do you believe that all values can be expressed in economic terms?

Values and worldviews

In the previous section, we explored some of the ways in which values underlie arguments and methodologies in economics and science. Yet values are embedded in broader worldviews as well. For example, many environmental philosophers have argued that our dominant contemporary worldviews are deeply *anthropocentric*: they are human-centered in the sense that they make humans the only or primary subjects of moral value and moral obligation. According to an anthropocentric worldview, animals and plants are not the direct bearers of moral value; instead they have value only in relation to their importance to human beings. From this point of view, we may have reason to protect the waters that support trout, but not for trout's sake. Instead, the reason to protect trout and their habitat derives from human interests, values, and needs. If people depend on trout for sustenance, enjoy fishing for trout, or appreciate the opportunity to see trout in clear mountain streams, then for those reasons, we ought to protect trout. Notice, though, that this leaves certain parts of the natural world vulnerable: for if there are parts of nature that do not matter to humans and on which we do not depend, anthropocentrism allows that there may be no reason not to damage, degrade, or destroy them. In Chapter 3, we examine various responses to anthropocentrism. Some of these approaches extend moral consideration to animals, others include all living things, and still others suggest that not only individual organisms, but also entities such as populations, communities, species, and ecosystems have moral value. According to these broader theories of moral value, we have direct obligations not only to other human beings, but also to other living things or other elements of the natural world.

For now, however, it will be helpful to think more carefully not just about anthropocentrism, but about how our worldviews and environmental attitudes, more generally, are shaped by important historical developments. These developments include the rise of mechanism in the seventeenth century, the development of Enlightenment individualism in the seventeenth and eighteenth centuries, the introduction of Darwinian evolutionary theory in the late 1800s, and the advent of ecology as a distinct scientific discipline in the twentieth century and beyond. This list of critical historical developments is not intended to be comprehensive, but illustrative. Reflection on these examples is intended to provoke more thoughtful consideration of the ways in which various social, political, economic, cultural, and scientific developments interact with our understanding of and outlook on the world and our place in it. It is also worth noting that although it is true that new ideas and approaches—political, institutional, technological, and so on—to some extent displace those that came before, our ways of thinking about and interacting with the world are shaped by many different layers of ideas and practices.

Even though you may know very little about Plato or Aristotle, for example, your worldview may have been shaped significantly by these thinkers, whose ideas are complexly embedded in contemporary life. Western thought has also been substantially shaped by Judeo-Christian religious traditions as well as by secular thinkers, such as the utilitarian philosophers Jeremy Bentham and John Stuart Mill. Although rarely acknowledged, Asian religious and philosophical traditions such as Buddhism, Hinduism, and Confucianism have also influenced the views of many Western thinkers, including such paradigmatic American writers as Henry David Thoreau, Walt Whitman, and

Ralph Waldo Emerson. The developments discussed below have affected the Western relationship to nature in particularly significant ways. The discussion aims to illustrate how various historical developments—though seemingly distant from us in time or space—nevertheless form an important context for our ways of thinking and acting in relation to nature today.

Mechanism and the Scientific Revolution

During ancient and medieval times, Western views of nature were grounded in a sense of natural order. Building on ancient Greek ideas, for example, the medieval philosophers described a "Great Chain of Being," in which nature was hierarchically arranged, with God at the top, followed by the angels, then humans, animals, and plants. The chain conceived a natural order in which each entity had a place and role in a larger organic unity.

Environmental historian Carolyn Merchant (1989, 1992) argues that this organic view of the world prevailed in Western thought until the Scientific Revolution. "[F]or sixteenth century Europeans [prior to the Scientific Revolution], the root metaphor binding together the self, society, and the cosmos was that of an organism" (Merchant 1989, p. 1). Merchant further suggests that a particular version of the organic metaphor was key: the metaphor of earth as mother. This metaphor not only provided an image of nature, but also guided action in relation to it. As a nurturing maternal figure, nature was comprehended as a living being worthy of respect and deference. In addition, certain activities, such as mining, were constrained by the conception of earth as mother. As Merchant explains, "One does not readily slay a mother, dig into her entrails for gold or mutilate her body . . . as long as the earth was considered to be alive and sensitive, it was considered a breach of human ethical behavior to carry out destructive acts against it" (Merchant 1989, p. 3).

Merchant argues that the rise of the mechanistic worldview in seventeenth century Europe represented a significant departure from prior modes of thought, a departure that enabled new ways of using—and abusing—the natural world. With the development of modern science, the organic worldview began to wane as a mechanistic worldview emerged in its place. This mechanistic conception emphasized the metaphor of earth as machine—and machines are importantly different from mothers. Rather than a living, nurturing, life-giving being, the earth was now seen as a collection of nonliving parts subject to human manipulation. Chemist Robert Boyle (1627–1692), for example, described nature as a giant clock (Taliaferro 2001, p. 134). The mechanistic view emphasized the idea that nature could be understood and controlled through division and dissection into smaller and smaller parts, tiny bits of matter that, "like the parts of machines, were dead, passive, and inert" (Merchant 1992, p. 49).

The mechanical model of nature brought with it a wide range of consequences. The nature-as-machine metaphor applied not only to the natural world as a whole, but also to individual living things. Philosophers such as René Descartes, for example, argued that animals could be understood as machines, lacking consciousness or rational capacities. This view in turn provided a way to justify vivisection. As Colin Allen (2011) explains, "Whether or not Descartes himself practiced vivisection (his own words indicate that he did), the mechanists who followed him used Descartes' denial of reason and a soul to animals as a rationale for their belief that live animals felt nothing under their knives."

There is not, of course, a necessary connection between a mechanistic, reductionistic approach to understanding the natural world and the aim of dominating or manipulating that world. What the nature-as-machine metaphor facilitates is the application of our moral frameworks for machines to those entities portrayed as machines, just as the nature-as-mother metaphor facilitates the application of our moral thinking about mothers to nature. This shows that metaphors can be very powerful in shaping our relationship to the natural world. However, they can also be critically examined, as can the normative assumptions they tend to carry with them. This may prompt us to craft better metaphors, or to modify our assumptions about what our obligations—to mothers, machines, or other entities that figure in nature metaphors—really require.

Enlightenment individualism

The mechanistic philosophy of the Scientific Revolution not only provided a way of approaching scientific inquiry, but also informed political philosophy and human self-understandings. Mechanism and reductionism assumed that entities were composed of basic parts and could be understood through an examination of these parts, which would reveal fundamental laws governing their behavior. These views were applied not only to animals, plants, or other elements of nonhuman nature, but to human beings as well (though in a slightly tempered form). In particular, the conception of the individual that emerged during the Enlightenment represented persons as rational, autonomous, and self-interested, and society as a collection of individuals bound to one another by contractual relations based on rational consent. Although the cover of Thomas Hobbes' major work, *Leviathan*, depicts society as a single organism—a man—one can see that this man comprises many smaller parts, and these parts are themselves individual persons. Individuals are subject to the rules and restrictions of the Leviathan, or sovereign ruler, but they remain independent, atomic individuals, free to pursue their individual desires and goals within the confines of certain broad constraints.

Robert Goodin (1998, p. 531) argues that the view of the self as autonomous and independent "has reigned supreme throughout mainstream Western moral and political thought." The view is well illustrated in the following quotation by Pico della Mirandola, who puts these words, addressed to the biblical Adam, into the mouth of God:

> You shall determine your own nature without constraint from any barrier, by means of the freedom to whose power I have entrusted you . . . so that like a free and sovereign artificer you might mould and fashion yourself into that form you yourself shall have chosen. (Pico della Mirandola, "Oration on the dignity of man," quoted in Goodin 1998, p. 531)

For our purposes it is important to note that according to this view, the individual exists prior to society, and society exists precisely in order to further the individual purposes of its members. These ideas may seem very natural to us today, as they form the basis for political liberalism, the dominant political philosophy of the United States and Europe, which emphasizes freedom and equality. At the core of the liberal view is the idea that individuals should be free to pursue their own conceptions of the good insofar as these pursuits do not interfere excessively with others' ability to do likewise. In other words, under liberalism the primary justification for limitations on individual liberties derives from the ideal of liberty itself: we can restrict individual freedom to act in cases where the actions in question would deprive others of their freedoms. For example, although freedom of speech is highly prized in the United States, it is not legally permissible to intentionally incite violence through speech.

There is much to be said in favor of the liberal point of view, yet it is not without its limitations. In particular, due to its focus on individual rights, liberalism may make it difficult to secure certain collective goods. As we will see later in this book, the conception of individuals as self-interested, rational actors can generate collective action problems such as the tragedy of the commons. A tragedy of the commons occurs when it is the interest of each individual to utilize an ever-increasing share of the commons, yet the pursuit of this interest by each individual results in the collapse of the commons as a whole. In the classic case described by Garrett Hardin (1968), we are asked to imagine a shared field where farmers graze their sheep. Each farmer has an incentive to add another sheep to the field so that he may raise and bring to market as many sheep as possible, but if individuals add sheep without restraint, the field as a whole will become denuded and incapable of sustaining grazing animals at all.

As we move to a discussion of cost–benefit analysis in Chapter 2, and to specific environmental problems, such as climate change, in later chapters of this book, it will be useful to consider the ways in which the Enlightenment conception of the self informs and shapes our thinking about these issues. We will also see that there are ways in which the dominant Western view has been challenged. The Western tradition is far from monolithic, and objectors to strong individualism have argued for alternative conceptions of the self that emphasize relationships and interdependence. What's more, not all traditions put such a strong emphasis on freedom and autonomy. The Confucian tradition, for example, stresses the individual's role in a larger social order, and indeed sees as incoherent the idea that a person can truly exist in the absence of relationships with others. For Confucians, our relationships with family, friends, coworkers, and others play a critical, constitutive role in our identities.

Darwin and the theory of evolution by natural selection

The discussion of relationships and interdependence at the close of the last section connects well to certain aspects of Darwinian thought. For although one reading of Charles Darwin's theory of evolution by natural selection emphasizes "the struggle for existence," in which individuals compete with one another for opportunity to survive and reproduce, it is also the case that Darwin's ideas challenge Pico della Mirandola's notion that humans are a unique creation of God whose characteristics differ radically from those of all other creatures. Pico's God says, "I have made you neither heavenly nor earthly, neither mortal nor immortal" (quoted in Goodin 1998, p. 531), and he asserts that it is this that allows humans the freedom to be fully self-determining. While Darwin does not directly inveigh on the question of human freedom, his work does cut against human exceptionalism of this kind. We are tied, through a great tree of evolutionary relationships, to all other living things, Darwin argues. Darwin's theory of evolution provided a mechanism—natural selection—by which species change over the course of generations, and through which the process of speciation, or the development of new species, occurs (Darwin 2001). In Darwin's view, evolution generally works incrementally, but through long stretches of time incremental changes can produce radical differences between ancestors and their descendants.

Natural selection is the process by which these evolutionary changes occur. Natural selection depends on variation among the individuals of a particular species, heritability (so that offspring tend to resemble their parents), as well as differences in survival and reproduction among different variants. For example, imagine a population of deer among which some are faster runners than others. These faster runners, we can suppose, are in general better able to elude predators than their slower counterparts. Consequently, the faster runners are more likely to survive and reproduce, and if speed is heritable, their offspring are likely to be rapid runners as well, and so more likely to survive and reproduce themselves. Over time, this process will produce a population of deer that run, on average, significantly faster than their forbearers.

The Darwinian perspective has been interpreted in various ways, with some emphasizing the competitive struggle for existence and "survival of the fittest" as a reflection of or support for the idea of free market economic systems, in which competition among businesses weeds out the weak and only the strongest and best survive (cf. Radick 2003). This perspective was also used by Social Darwinists of the late nineteenth and early twentieth centuries to suggest that public support of the poor and less advantaged was not justified because it interfered with the progressive process of natural selection that would lead to improvement of the human race over time. Unsurprisingly, these ideas were controversial. The case of human evolution is complicated substantially by culture and by the fact that we have a significant role in constructing our own environments, shaping the selective pressures we encounter. What's more, many philosophers of biology have emphasized that evolution by natural selection does not necessarily lead to progress in an unqualified sense (for discussion, see Rosenberg and McShea 2008, ch. 5). Rather, natural selection generates, over time, a better "fit" between organisms and their environments; if environments change, then traits selected for previously may no longer be adaptive.

For our purposes, the critical point that emerges from Darwin's theory is the idea that human beings—though exceptional in certain ways—share a lineage with other living things, are in fact recent arrivals on the planet, and have skills that exceed those of other organisms in some respects, but are dwarfed by them in others. The theory of evolution by natural selection implies that diverse species are related to one another and have developed through fundamentally similar processes. Moreover, evolutionary biology shows that human beings are a fairly recent addition to the planet. While the history of life on earth extends back about 3.7 billion years, our species—*Homo sapiens*—is only about 200,000 years old. Lastly, while humans have certain characteristics—advanced language capacities, complex and robust culture, and impressive capacities for abstract thought—that distinguish them from other animals, we lack the speed of a cheetah, the keen scent of canines, or the echolocation abilities of bats and dolphins. For environmental philosophy, these ideas are important because they suggest a kind of Copernican revolution for human beings: they destabilize the view that human beings are at the center of the world, and that everything else exists for our sake. As environmental philosopher Paul Taylor (1981, p. 207) put it:

> When we look at ourselves from the evolutionary point of view, we see that not only are we very recent arrivals on Earth, but that our emergence as a new species on the planet was originally an event of no particular importance to the entire scheme of things. The Earth was teeming with life long before we appeared . . . [W]e are relative newcomers, entering a home that has been the residence of others for hundreds of millions of years, a home that must now be shared by all of us together.

We shall see when we discuss animal ethics that an evolutionary perspective can help support challenges to the claim that human beings are categorically different from other animals. Many of the traits we have used to mark ourselves as exceptional—such as language, rationality, consciousness, emotion, and culture—are present to some degree in other species as well. Of course, the softening of the boundaries between humans and other animals can cut in two directions, with some arguing that if we are just like other animals, then we are no more morally responsible for our behavior than they. If animals aggressively kill and eat other animals, without concern for their suffering, then why shouldn't we? This is a question worthy of thought and discussion—but it is important to note at this point that even those who argue for continuity between humans and other animals are not suggesting that we are exactly the same, or that animals either deserve exactly the same treatment or should be held to exactly the same standards as human beings. As Peter Singer (2008, p. 74), who argues that we should weigh equally the interests of all beings (including animals) who have interests, explains, "The extension of the basic principle of equality from one group to another does not imply that we must treat both groups in exactly the same way, or grant exactly the same rights to both groups."

Ecology, interconnection, and the "balance of nature"

Although Darwin's theory of evolution emphasizes our connections with other living things through shared ancestry, the science of ecology emphasizes the ways in which diverse organisms are connected to one another through their interactions as elements of larger ecological systems. The science of ecology emerged as a distinct discipline early in the twentieth century, growing from a much longer tradition of natural history observation and biological theorizing. Ecology—derived from the Greek words *oikos*, meaning "home," and *logos*, referring to fundamental principles—focuses on the relationship between organisms and their environments. Ecologists typically think in terms of hierarchical levels of organization, ranging from individual organisms to populations to communities to ecosystems. This hierarchical approach recognizes the importance of various scales in generating ecological patterns and processes. Ecology is not, strictly speaking, a reductionist science, though the approach of understanding a system by examining its parts and their functions is certainly not alien to the discipline. Nevertheless, by keeping multiple levels of organization in view, ecology cuts against the idea that every legitimate scientific question can be answered by dissecting the natural world into its smallest parts. Ecology thus gives greater recognition and ontological status to entities at higher levels of organization than does a discipline like physics.

This last point is significant for environmental philosophy, and particularly for arguments in environmental ethics that aim to establish moral standing not only for individual organisms, but for entities such as populations, species, ecological communities, and ecosystems. For if ecological communities are nothing more than collections of individual organisms of various kinds that happen to occupy the same geographic area, then it may be difficult to arrive at a justification for protecting or valuing communities per se. In such a case, it would seem that there would be little reason to recognize the community itself, as distinct from the particular, individual organisms that comprise it, as having value. Thus, by validating the existence of larger ecological wholes such as communities and ecosystems, ecological science opens up the possibility that these wholes might themselves be worthy of moral consideration.

A second way in which ecology has provided an important context for discussions in environmental philosophy relates to the balance of nature view that was central to ecology throughout much of the twentieth century. Although this view takes various forms, the basic idea was that there exists a balance of nature in the sense that natural systems tend to be relatively stable over time, with variation occurring within a relatively small range (Cooper 2003, pp. 76–77). Early twentieth century ecologist Frederic Clements captured a related view of the balance idea in his model of plant succession. According to Clements, multispecies plant communities could be understood on the model of a superorganism, following a predictable pattern of development from youth to maturity. Clements held that individual plant species have a function in succession just as individual organs have a function in the development of whole organisms. In this model, once a plant community reached its mature, or climax, state, it would remain in equilibrium in the absence of significant disturbance.

The balance of nature idea thus suggested that if ecological systems were allowed to develop unimpeded, they would reach a steady state. In turn, it was often assumed that nature would remain stable and balanced if humans would just leave it undisturbed. The concern with this view, some have argued, is that it seems to leave no constructive role for humans in relation to the natural world, and this is problematic, for we need to learn to live and act well in and with nature (see Cronon 1995).

The balance of nature view, however, has been at least partially dislodged. In the last few decades of the twentieth century, ecologists began to recognize the crucial roles that disturbances play in the overall functioning of ecological systems. Many forests, for example, developed in association with natural or human-caused fire. In *Discordant Harmonies*, ecologist Daniel Botkin (1990) explained that the Hutcheson Memorial Forest Nature Preserve in New Jersey, an oak-hickory forest originally believed to be the product of "undisturbed nature," actually developed as a result of Native American burning practices that eliminated sugar maple seedlings that would have otherwise come to dominate. Botkin (1996) further argued that the balance of nature paradigm obscures the role of human beings in creating ecosystems we value today and discourages the careful investigation of ecological processes needed for effective management. Disturbance—whether by fire, wind, flooding, or other processes—gradually has come to be recognized as much more important in shaping natural ecosystems than was traditionally believed. Relatedly, we now recognize problems and oversimplifications in the idea that ecological systems develop in a steady, predictable progression to a climax state. Botkin's conclusion—that "we need to accept the variation in nature so that we can move toward a constructive, active role for people in Nature, and become able to formulate policies, based on Nature's variations, that work" (Botkin 1996, p. 38)—is gradually being taken to heart.

Questions about the possibility of finding an active, yet constructive role for humans in nature are particularly pressing today, as we grapple with the consequences of large-scale human-induced changes to the environment, from habitat destruction and species extinctions to climate change, with its attendant consequences for sea level, weather patterns, and terrestrial, freshwater, and marine ecosystems. Nature is in a constant state of flux, and humans have always altered their environments, but the scale and magnitude of the changes we are now generating is unprecedented in the lifetime of our species. What's more, the current rate of climate warming, catalyzed by anthropogenic emissions, appears to exceed that of any other warming in the past 1 million years (IPCC 2007a). Nobel Prize-winning atmospheric chemist Paul Crutzen, among others, have gone so far as to dub our current geologic era the Anthropocene, or the age of humans. As Paul Crutzen and Christian Schwägerl (2011) put it, "the Anthropocene—human dominance of biological, chemical and geological processes on Earth—is already an undeniable reality." Although the truth of this descriptive claim about human dominance entails no specific conclusions about how we ought to act, many of the Anthropocene's strongest boosters argue that we should acknowledge and embrace the power and control we now exert over our home planet. Erle Ellis (2011), for example, writes:

> The Earth we have inherited from our ancestors is now our responsibility . . . [T]here is no alternative except to shoulder the mantle of planetary stewardship. A good, or at least a better, Anthropocene is within our grasp. Creating that future will mean going beyond fears of transgressing natural limits and nostalgic hopes of returning to some pastoral or pristine era.

Notice how this recommendation explicitly contradicts the idea associated with the balance of nature view that "nature knows best." Ellis insists that it is now up to us to determine what is best. This, of course, raises a host of difficult questions, especially if we are talking about large-scale environmental issues such as global climate change. If we embrace Ellis's recommendation, then who determines what is best, and how should they do it?

For further thought

1. What are the strengths and limitations of (a) the nature-as-mother metaphor and (b) the nature-as-machine metaphor? Choose another possible metaphor for nature and consider its potential ethical implications.
2. Identify and describe a historical, cultural, or religious perspective other than those described above that has significantly shaped your worldview. How might this perspective be relevant to thinking about the environment?
3. Does accepting a role for regular disturbances in ecological systems imply that there are no natural equilibria, and no natural balances of any kind?
4. Can human beings take an active role in managing natural environments without dominating them? Why/why not?

Ethics, metaethics, and moral progress

The question just posed—who determines what is best?—brings into focus a general challenge for environmental ethics. If environmental ethics is a subject that concerns how we ought to act in relation to the natural world, then how do we go about determining this? For that matter, is there really a "right answer" to questions in environmental ethics? Is there a fact of the matter as to whether we ought to protect tigers from extinction? About whether we should leave some parts of the world undeveloped and free of human habitation? About whether we should or should not attempt to intentionally manipulate the climate to counteract the effects of global warming?

The questions raised in the previous paragraph concern both metaethics and moral epistemology. *Metaethics* is the area of ethics that concerns the nature and foundations of value, and the meaning of moral terms. Whereas *normative ethics* deals with how we ought to live and what makes an act right or wrong, metaethics is concerned with the nature of morality more generally. Metaethical questions include those about whether morality is objective, subjective, or culturally relative, and about what we mean when we make moral claims such as "murder is wrong." Questions concerning *moral epistemology*, or the nature of moral knowledge, also fall within the broad purview of metaethics. Moral epistemology involves questions about how we attain moral knowledge and what counts as sufficient justification for moral beliefs.

Metaethics can seem abstract, but metaethical questions arise with surprising frequency in everyday life. For example, common moral disagreements raise not only normative questions (Is lying always wrong?), but metaethical questions about whether there even is a single, correct answer to the question of whether lying is always wrong. Let us take an example to illustrate. Imagine that you think that even if humans could live decent lives on a planet with many fewer and many different species than exist now, it would be wrong to drive species to extinction, and to remake the earth in ways that suit us as human beings at other species' expense. But you have a friend who feels differently. She thinks that what is most important, morally speaking, is to make the planet a hospitable place for humans to live. Although she acknowledges the importance of maintaining functional ecosystems insofar as they benefit human beings in numerous ways, she sees no reason to sacrifice progress toward greater human well-being for the sake of other species on which our existence does not depend.

What should we say about this disagreement? Can it be rationally adjudicated? Or are these merely differences of opinion, that no rational argument can decide?

Subjectivism, relativism, and universalist realism about morality

If you believe that the answers to ethical questions are determined by or dependent on each individual's moral beliefs, then you are a subjectivist.

In response to the disagreement, a subjectivist might say that it is right for the environmentalist to work to save species from extinction, but that the environmentalist's friend has no obligation to do so. *Subjectivism* is a metaethical position that holds that right and wrong are indexed to (or relative to) a particular subject's moral beliefs. Notice that this view seems to make very little space for rational discussion among individuals regarding controversial ethical issues, or for a person to be wrong about what he or she is morally obligated to do. According to the subjectivist, I can be mistaken about whether I ought to eat meat, for example, only if I am confused about the implications of my own moral commitments. This can, of course, be the case, and some authors, such as Mylan Engel (2000) have argued for vegetarianism by attempting to show that each of us already has individual moral commitments that entail that we refrain from eating meat. However, the subjectivist must concede that if a person lacks the relevant commitments, then that person lacks the corresponding obligation. According to moral subjectivism, there is no basis for asserting obligations that apply to all persons, regardless of their subjective values.

Moral relativism has a structure that parallels that of subjectivism, but in this case, right and wrong are not indexed to particular individuals, but instead to larger social groups. Christopher Gowans (2012) defines metaethical moral relativism thus:

> The truth or falsity of moral judgments, or their justification, is not absolute or universal, but is relative to the traditions, convictions, or practices of a group of persons.

According to the relativist, whether vegetarianism is right or wrong for me is relative to the beliefs and practices of my society or culture. Relativism allows for the possibility that it may be right for people in one social group to eat meat while it is wrong for people in another group to do so. Because relativism has structural similarities to subjectivism, it suffers from some of the same kinds of difficulties. For example, one might worry that relativism precludes constructive cross-cultural moral dialogue, for it makes a group's beliefs determinative of right and wrong. However, there are some more subtle and complex forms of relativism that allow for common universal constraints on moral systems (see Wong 2006), and relativisms of this form may better allow for substantive moral discussion among different cultural and social groups.

A third key metaethical position is *universalist moral realism*, which holds that there is a single, true morality, invariant across time and space. What is right or wrong to do may of course depend on the circumstances or context—for example, universalism is compatible with the possibility that something that is generally wrong (like lying) may under certain conditions be morally permissible. Nevertheless, right and wrong are not dependent on what a particular person or group of people believes to be right and wrong. For environmental philosophers, this may be encouraging—because a metaethical view that allows for some distance between what people believe to be right and what actually is right seems to make room for the possibility of moral progress. Environmental philosophers frequently challenge the status quo, and in doing so, they sometimes recommend radical changes in our ways of valuing and interacting with the natural world. Universalist moral realism, in holding that moral truths are objective and independent of people's actual moral beliefs, allows for the possibility that the anthropocentrism, though widespread, may be wrong. Thus, perhaps, a minority view like biocentrism—which affirms that all living things are worthy of moral consideration—may be right.

Nevertheless, moral universalism—despite its appeal in this regard—faces several challenges of its own. Some of these challenges are epistemological: if there is a single true morality, how can we come to know it? In addition, moral realists face pressure to respond to the observations that motivate moral relativism: there exists a substantial amount of moral disagreement in the world, and some of these disagreements seem not to be rationally resolvable. What's more, moral realism is not the only metaethical view that allows for the possibility of moral progress.

Reflective equilibrium

For the purposes of this book, we do not need to settle the metaethical questions just discussed, but it will be helpful to be aware of these positions and their implications. This is particularly the case because we want to allow for the possibility of moral criticism, constructive dialogue, and moral improvement, and some forms of subjectivism and relativism seem to limit progress in this regard. If we think that each person's (or each group's) current moral beliefs settle what is right or wrong, at least for the person (or group) in question, then how do we go about working through our disagreements?

Regardless of where one ultimately comes down on the metaethical questions, we need a general strategy for approaching and working through controversial ethical issues. Even if we diverge in our specific ethical views, most of us share the desire to live in a relatively harmonious and well-coordinated society, which requires coming to decisions on a number of things, such as how and to what extent pollution should be regulated, how many lands to set aside from development, how various lands should be managed, and so on. As should be clear by now, these are not just scientific, economic, and policy questions, they are also ethical issues. But if we do not know whether there exists a single true morality, or what its content might be, how do we go about settling such questions?

Even if there is no single true morality, some ethical views may be better than others. As individuals and as a society, we can sift through various alternatives and attempt to identify the best from among them. It may be helpful here to take the analogy of a literary text. With any complex text, there is more than one viable interpretation. Nevertheless, there are some interpretations that are implausible and which can be ruled out. Similarly, excellence in a particular sport or musical instrument cannot be evaluated with a single, simple formula, but there are some common factors that characterize a good athlete, or a good musician. One might expect something similar in the realm of ethics: we might never find a simple algorithm for discovering moral truth, but we can nevertheless draw on various resources in identifying viable ethical positions.

Inevitably, we will need to start—to some degree—from where we are. The method of reflective equilibrium, developed in part by the American political philosopher John Rawls (1971), can help in this regard. Rawls's approach uses reflection to bring ethical principles, broader worldviews, and intuitions about specific cases into alignment with one another. The idea is that when one's principles conflict with one's ethical beliefs about specific cases, for example, one must make adjustments, either to the principles or to more specific ethical beliefs in order to bring them into balance.

An example can help illustrate this idea. Assume that I am committed to the view that pain is, in general, a bad thing. I believe that all else equal, a world with less pain is better than a world with more of it. Yet, I am also a regular consumer of items whose production involves animal suffering, even though I do not need these things to survive or flourish. Using the method of reflective equilibrium, I examine these potentially conflicting commitments. In doing so, I might find that my belief and my practices are not inconsistent, because although I believe that pain is a bad thing in general, I think that animal pain and pleasure count for less than human pain and pleasure. In this case, I determine, the pleasure I gain from using various animal products outweighs the suffering involved.

Two things are to be noticed here. First, the method of reflective equilibrium can be relatively conservative. It pushes us toward coherence among our beliefs and between beliefs and practices, but it offers no new foundational principles that must displace the beliefs one already holds. However, the method of reflective equilibrium can bring to light inconsistencies and incoherence in our moral commitments, and so generate pressure for change. In certain cases, the pressure for change can be quite significant. In the example above, although I find that my commitments are not inconsistent, I have been prompted to articulate why they are not, and this process of making explicit my reasons may open up new possibilities for ethical reflection and change. If, for example, I enter into conversation with friends and articulate my reasoning to them, I may be pressed to explain the ethical commitment that allowed me to bring my apparently conflicting commitments into alignment: Why is it that I believe that human pleasure and pain count for more than animal pleasure and pain? Although I may have a good answer to this question, one can see how the approach can prompt us to examine and justify our views in a more thorough and systematic way. Just as careful analysis of the assumptions involved in science and economics can lay bare value commitments embedded in these disciplines, the method of reflective equilibrium helps us lay bare the values that underlie our own reasoning and action.

Ethics and the connected critic

We have spent much of this chapter considering the ways in which values are embedded in science, economics, and our worldviews, more generally. The discussion has aimed to highlight the way in which ethical reflection always begins *in medias res*, or in the middle of things. Each of us has ethical convictions that we bring to environmental ethics, and our individual views are embedded in a larger social context, which itself embodies various moral values. In the last section of the chapter, we discussed the question of whether an individual's or a society's values are morally determinative, or whether what is ethically right might be independent of what any particular person or society believes to be right. The method of reflective equilibrium, we have noted, carries with it an inherent conservatism: it begins with beliefs held by an individual or society and seeks coherence among those beliefs, which allows for change, but insists at the same time on continuity. Radical change—involving a wholesale shift in beliefs—seems to be unlikely, or even impossible, using this method. Yet, *significant* change is possible. The method of reflective equilibrium can, for example, reveal to us the prejudicial application of our moral principles and encourage us to expand their scope. What's more, when we employ the method as part of a process of social dialogue, we must confront not only our own beliefs and intuitions, but also those of others. Finally, a thoroughgoing reflective process must consider not only the particular moral beliefs and intuitions we have, but how they fit into the larger background theories about the world and our place in it. If we engage in this broader reflective process, we can reach not just a narrow reflective equilibrium, but what John Rawls calls wide reflective equilibrium. As Norman Daniels (2011) explains,

> [Rawls] comments that seeking a reflective equilibrium that merely irons out minor incoherence in a person's system of beliefs is not really the use of the method that is of true philosophical interest in ethics.

This narrow approach stands in contrast to wide reflective equilibrium, which involves "searching deliberation about what is right. It is this much broader [form] of challenge that Rawls labels the method of wide reflective equilibrium" (Daniels 2011).

On this view, wide reflective equilibrium lays open the possibility of significant change and significant moral progress. In environmental ethics, wide reflective equilibrium will have to take into account challenges from a broad range of perspectives, including views such as biocentrism, which challenge the status quo. Nevertheless, we always begin the process from a particular ethical location, embedded as we are in societies and traditions that help define for us a way of life and constellation of moral values.

It remains an open question whether the method of reflective equilibrium will be sufficient to enable us to achieve an adequate environmental ethic. In Chapter 3, we discuss theoretical positions such as deep ecology that may require a wholesale "paradigm shift" in our way of understanding ourselves and our relation to the natural world. If such a shift is in fact needed, how might we achieve it? From a practical perspective, a radical break with our current ways of thinking seems to require something more like conversion rather than rational persuasion. Rational persuasion involves building from shared premises, and social criticism in this vein may fit best with what Michael Walzer (1987) calls the model of "the connected critic." Walzer describes this critic as one who consistently engages with others in the community through questioning and arguing. This critic is one who "objects, protests, and remonstrates," but is not alienated from the community: the connected critic is neither intellectually nor emotionally detached (Walzer 1987, p. 39). On Walzer's view, the effective critic is one who engages his or her fellow citizens in thoughtful, honest reflection, acknowledging and at the same time challenging prevailing views. As he explains, "[T]he connected critic . . . earns his authority, or fails to do so, by arguing with his fellows . . . This critic is one of us" (Walzer 1987, p. 39).

For further thought

1. Which metaethical position do you find most plausible: subjectivism, relativism, or universalist moral realism? Why?
2. Why might moral realism be particularly appealing to environmental philosophers critical of our current relationship to the natural world?
3. How might the method of reflective equilibrium lead to progress in environmental ethics?
4. Michael Walzer champions the model of the "connected critic." What are the strengths and limitations of this model?

Conclusion

This book aims to encourage reflection on the need and possibilities for change in our views about and our relation to animals, plants, ecosystems, and the environment, more broadly. In reflecting on these issues, it will be helpful to consider not only what you believe to be the strongest positions from a philosophical perspective, but also how our institutions and practices might better embody these views. For there are at least two key challenges we face in environmental ethics. In some cases, we will find deficiencies in our moral beliefs, which call for change and revision in these beliefs. In other cases, we will affirm our beliefs on reflection, but notice a significant gap between our beliefs and our practices. At a personal level, this gap may be as simple as believing that one should reduce wasteful consumption of resources, yet failing to turn out the lights when one leaves home. Although the purview of philosophical ethics traditionally has been restricted to investigating beliefs, this book will encourage you to think also about the relationship between beliefs and action. In Chapter 8, we return explicitly to these themes, considering anew the connections between environmental ethics and social change.

Further reading

Botkin, D. (1990). *Discordant Harmonies: A New Ecology for the Twenty-First Century.* New York: Oxford University Press.

Cooper, G. J. (2003). *The Science of the Struggle for Existence: On the Foundations of Ecology.* New York: Cambridge University Press.

Daniels, N. (2011). "Reflective equilibrium," in E. N. Zalta (ed.), *The Stanford Encyclopedia of Philosophy*, Spring 2011 edition, http://plato.stanford.edu/archives/spr2011/entries/reflective-equilibrium/

Darwin, C. (2001). *On the Origin of Species*. Cambridge: Harvard University Press.

Kitcher, P. (2001). *Science, Truth, and Democracy.* New York: Oxford University Press.

Kuhn, T. (1996). *The Structure of Scientific Revolutions*. Chicago: University of Chicago.

Longino, H. (1990). *Science as Social Knowledge*. Princeton: Princeton University Press.

Merchant, C. (1989). *The Death of Nature: Women, Ecology, and the Scientific Revolution*. New York: Harper & Row.

Shrader-Frechette, K. S. and McCoy, E. D. (1993). *Method in Ecology: Strategies for Conservation*. New York: Cambridge University Press.

Walzer, M. (1987). *Interpretation and Social Criticism*. Cambridge: Harvard University Press.

2

Classical ethical theories and the environment

Introduction

In Chapter 1, we considered the ubiquity of values—in science, economics, and in comprising our worldviews more generally. Often, values are implicit: they underlie our thinking and acting, our institutions and practices, yet they are unarticulated, or vague. This chapter turns to a more explicit discussion of ethical values, as embodied in three classical Western ethical theories: utilitarianism, Kantian ethics, and virtue ethics. Chapter 3—which focuses on nonanthropocentrism—will introduce two classical philosophies that originated in East Asia: Confucianism and Daoism.

Just as social, political, and religious beliefs and institutions form a background against which our values are shaped, so too do ethical theories. The field of environmental ethics has been defined, in part, as a response to the perceived inadequacy of classical approaches to ethics, though as we shall see, these approaches also provide some important conceptual resources for thinking about human relationships to the natural world. This chapter thus explores classical ethical theories for three reasons: first, because studying these theories helps us understand more clearly the way in which our own ethical values have been shaped and can sharpen our understanding of many contemporary moral disagreements; second, because we will want to see how these theories fall short with respect to our relationships to animals and the environment; and third, because we will want to explore the potential for modifying, building upon, or extending classical theories in ways that more adequately address the environment and our relationship to it.

Each of these three tasks is consonant with the approach to moral reform described in Chapter 1, where change builds and extends from existing beliefs and values. It is worth noting at the outset that although all environmental philosophies in some sense build on prior values and conceptual frameworks, some depart quite radically from traditional views. What's more, we need neither choose just one theory to guide our moral thinking, nor should we assume that all moral thought and action must be strongly theory-guided.

Before discussing specific ethical theories, it is important to clarify that the ethical theories discussed in this chapter are *normative* ethical theories (Figure 2.1). These theories attempt to answer questions such as "What makes an action right or wrong?" or "What sort of person should I be?" or "How can I live a good human life?" As noted in Chapter 1, normative ethics is concerned with how we ought to live and what makes an act right or wrong, and can be contrasted with metaethics, which explores the nature of morality more generally. Thus, normative ethical theories focus not so much on the nature of moral value or the status of moral truth as on questions of how we should act in the world. They extend beyond attempts to describe morality, moving into the prescriptive realm: normative ethical theories offer moral guidance. However, as we shall see, the relationship between normative moral theories and action is complex. For example, putting a theory into action requires more than the application of a simple algorithm. What's more, many philosophers think that ethics cannot be captured by a single theory (see, e.g., Brennan [1992] and Midgley [1996]). Rather, multiple theoretical perspectives are needed to understand fully the moral dimensions of a given situation. Even more radically, some argue that theories are largely irrelevant to acting well in the world because they can never capture the complexity of the moral landscape. Theories—by their nature—are simplifications, and

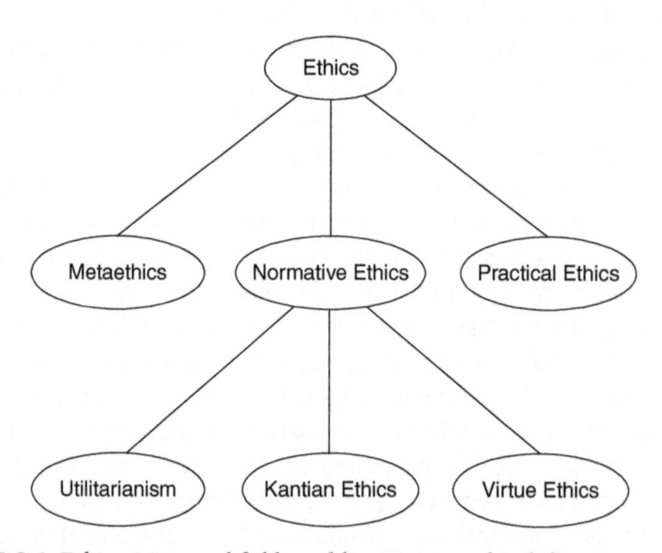

FIGURE 2.1 *Ethics: Major subfields and key Western ethical theories.*

moral theories work to clarify and systematize the basis for moral judgment and action by abstracting away from the details of particular situations. The challenge in relating theory to practice, then, is to determine how to integrate the simplicity of theory with the complexity of reality.

Utilitarianism, cost–benefit analysis, and the environment

Utilitarianism

Utilitarianism, like all moral theories, builds from certain key moral intuitions. In particular, utilitarianism draws on the intuition that *consequences matter* in ethics. This intuition is widely shared. In evaluating choices, we often consider the likely effects of these choices on ourselves and others and weigh these effects in our decision. To cite a famous example from philosopher Peter Singer: imagine that you were walking by a pond and noticed a small child drowning, just a few feet from shore. By wading in to save the child, you would get your feet wet, but a life would be saved. What should you do? Clearly, you should save the child, and most people reason that you should do so because the cost to you is minimal and the benefit to the child is great. The overall consequences for the world of saving the child are much better than the overall consequences of continuing on your way in order to save yourself the discomfort of wet feet. With respect to the environment, those who think that we ought to act to slow global climate change typically believe that we should do so because reducing greenhouse gas emissions will have significantly better consequences than failing to do so, even taking into account the fact that reducing greenhouse gas emissions itself will generate some negative consequences.

Interestingly, even those who oppose regulation of greenhouse gas emissions often do so on consequentialist grounds. They argue—rightly or wrongly—that the negative effects of climate protection regulations will outweigh the benefits. Those on both sides of the debate acknowledge the basic point that we should take seriously the consequences of various alternative courses of action. Moreover, those who favor regulation generally seek forms of regulation that would generate the greatest benefits while averting the worst harms.

Although many would agree that considering consequences is a critical dimension of environmental ethics, consequentialists believe that consequences are the *only* morally relevant factor that one should consider. Utilitarianism, as a form of consequentialism, shares this view. Virtue and vice, rights and responsibilities, or good and bad intentions are relevant only insofar as they generate good or bad consequences. From this perspective, questions like, "Should I tell the truth?" cannot be answered by appeal to some universal moral principle that prohibits lying. Instead, consequentialists ask, "Well, what are the consequences of telling the truth, and what are the consequences of lying? Compare those, and you'll have an answer to your question." Although the view that *only* consequences matter morally may seem unfamiliar—surely, honesty and courage are good in themselves, one might think—a consequentialist might counter by asking whether it is not the case that we value courage because of the good consequences it produces. If courage generally created more suffering than it prevented, would we still value courage?

As noted above, utilitarianism is a consequentialist moral theory, a theory that holds that good consequences are the sole moral goods. In particular, utilitarians hold that what one ought to do, or what is morally right, is that which *maximizes good consequences* (and minimizes the bad). Utilitarianism derives its name from the idea of British philosopher Jeremy Bentham (1748–1832) that the fundamental principle of morality is to maximize utility, or happiness. Bentham understood utility as pleasure and the absence of pain; thus Bentham's version of utilitarianism is known as *hedonistic* utilitarianism (from the Greek *hēdonē*, meaning "pleasure").

Bentham's utilitarian theory is attractive in its egalitarianism: everyone's pleasures count, whether simple or sophisticated, humble or elite. As Bentham puts it, "Pushpin [a simple child's game] is as good as poetry." While this neutrality can have significant appeal, it was criticized by Bentham's successor John Stuart Mill as well as many contemporary philosophers. Surely, the pleasure that one person takes in shooting small animals for fun should not be equated with a similar amount of pleasure that another person feels in helping others. Benthamite utilitarianism accounts for the negative effects on overall happiness of the first activity and the likely positive effects of the second, but even so, many people have the intuition that pleasure taken in shooting animals should *itself* be discounted relative to the pleasure of assisting others.

In addition to hedonistic utilitarianism, where utility is understood in terms of psychological states—pleasure and the absence of pain—contemporary philosophers such as R. M. Hare have advanced an alternative version of the theory known as *preference utilitarianism*. According to this view, utility is not measured in terms of pleasure and the absence of pain, but instead by the satisfaction of preferences. One key strength of this view is that it can help overcome objections to hedonistic utilitarianism that question whether pleasure is really the only good with which we should be concerned. The philosopher Robert Nozick (1974, pp. 42–45) puts this objection in the form of a thought experiment. He asks us to consider whether it would be better to live in a *Matrix*-like scenario, attached to an "experience machine" that provides a great deal of pleasure based on the illusion of caring friends, a loving family, a successful career, and so on, than to live in the real world that includes many pleasures, but also a share of setbacks and disappointments. If offered the opportunity to hook yourself up to the machine, Nozick asks, would you do it? If not, then pleasure and the absence of pain are not the only goods that concern you. Many people, for example, would rather have real friends—and the real challenges that accompany these relationships—than the illusion of friends, even if these latter "friends" are the source of great pleasure. Preference utilitarianism can accommodate this intuition, taking seriously the idea that the best kind of world is not necessarily one in which pleasure is maximized, but one in which the satisfaction of preferences is maximized, even if that latter world is not the one with the most pleasure.

In any case, utilitarianism in each of these basic forms neither makes qualitative distinctions between pleasures nor passes judgment on preferences. This avoids the need to adjudicate among competing visions of the good life, and indeed, this is one reason why utilitarianism fits so well with political liberalism. Liberalism, the dominant political philosophy of the Western world, holds that each person should be free to pursue his or her own conception of a good life, so long as doing so does not impede others from doing the same. Whether neutrality about preferences is a good thing will be discussed in the following section—some critics contend that this is actually a significant flaw in the utilitarian approach and its cousin, cost–benefit analysis (CBA).

Before turning to CBA, however, let us consider some of the key strengths of utilitarianism. We have already noted that utilitarianism captures the widespread moral intuition that consequences matter morally. Extending this point, we might see as a strength the fact that utilitarianism stresses our obligation to make the world a better place by acting to maximize positive consequences. Second, the theoretical simplicity of utilitarianism is appealing: the theory first defines the good (typically as either pleasure or preference satisfaction), then tells us to maximize it. The right thing to do is defined very clearly: maximize utility! Third, and relatedly, utilitarianism provides a means to commensurate values, to bring them together in a single scale. If I take pleasure in playing football, and you take pleasure in playing the cello, we need not adjudicate this difference in taste. Instead, utilitarianism simply asks us to calculate the effects of alternative actions on the amount of overall pleasure in the world, taking everyone's pleasures into account.[1] The theory remains neutral regarding how pleasure is produced. By taking all pleasures and pains into account, utilitarianism aggregates apparently diverse values into a single measure of total utility. It neither weighs one individual's pleasure more strongly than another's, nor excludes any individual's pleasures and pains from consideration.

Two strengths of utilitarianism have special relevance to environmental ethics. The first of these concerns animals. Hedonistic utilitarians calculate utility by adding the sum total of pleasures and pains, wherever they occur. This means that we must not only consider the pleasures and pains of ourselves, or our friends and relatives, but also of *any* individual affected by our actions. What's more, according to Jeremy Bentham's original theory, the individuals in question are all those capable of feeling pleasure and pain. Since there is very good evidence that many animals are sentient, or capable of feeling pleasure and pain, their pleasure and pain should count just the same as that of humans. Although some philosophers contest the idea that animals can have preferences, it does seem that they *express* preferences—in choosing one kind of food over another, or protesting being put on a leash, for example. On this basis, it seems that a thoroughgoing preference utilitarian must take animal preferences into account.

The inclusion of animals in the moral calculus is a radical aspect of utilitarianism as compared to other traditional Western moral theories, virtually all of which exclude nonhuman animals either implicitly or explicitly. Thus, in theory at least, utilitarianism is substantially less anthropocentric than many of its classical rivals, and many environmental philosophers see this as a virtue. If we took seriously the sufferings of animals, for example, we might be led to transform our food production systems and approaches to animal experimentation. Factory farms might be abolished, and vastly fewer animals would be harmed or killed for trivial gains in knowledge (see Singer 2002).

One final strength of utilitarianism is that it translates easily into decision-making procedures, especially in the policy arena. By enabling apparently diverse values to be integrated and placed on a single scale, utilitarianism provides a systematic way to evaluate alternative environmental policies with diverse effects. In the following section we discuss cost-benefit analysis, which shares many basic structural features with utilitarianism, with the addition of assumptions that enable preferences to be translated into monetary terms.

Cost–benefit analysis and economic valuation

Cost–benefit analysis is a planning and decision-making tool whose structure parallels that of preference utilitarianism. The basic idea of CBA is to compare alternative laws, policies, or actions by adding up costs and benefits on a common scale. Following the utilitarian logic, we should then choose that alternative with the greatest net benefits. CBA mirrors the structure of utilitarianism, and many argue that it is the best strategy for making decisions that maximize overall utility (or welfare), understood in terms of preference satisfaction.

Although there are various ways to conduct a CBA, it is often assumed that costs and benefits, quantified monetarily, can serve as proxy measures for preference satisfaction. If this assumption holds, the alternative that has the greatest balance of benefits over costs will be that which maximizes preference satisfaction. A simple example can illustrate the basic strategy of CBA and the methods it uses to translate preferences into dollars. Imagine that the government is considering a new regulation to limit heavy metals in drinking water. The primary costs of these regulations will be those associated with adding filtration and monitoring systems for municipal water supplies. The main benefits will be better health for water consumers. The cost of adding filtration and monitoring can fairly easily be measured in dollars: How much will the new equipment cost? How much additional labor will be required to track levels of the heavy metals covered by the new regulation? The benefits of regulation in terms of better health are less straightforward to quantify, however, because health is not a market good: it is not bought and sold like consumer products such as cars, televisions, or food.

There are various ways to estimate benefits associated with nonmarket goods. Two common approaches are *hedonic pricing* (or "shadow pricing") and *contingent valuation* (or "willingness-to-pay"). Hedonic pricing works by comparing the costs of items that are similar in many respects but differ in relation to the feature one is attempting to price. If one wants to know, for example, the value of organic produce as compared to conventional produce, one could examine fruits of similar kinds, appearances, and freshness in both categories, looking at the size of the premium people pay for the organic label. Similarly, to determine the benefits of unpolluted air, one might compare the price of similar homes in similar neighborhoods, with different levels of air quality. If all other features of the homes are similar, hedonic pricing will attribute the price differential to differences in the cleanliness of the air. A second strategy relies not on actual market behavior but on people's stated preferences. Contingent valuation techniques use surveys to elicit the strength of people's preferences for various nonmarket goods. One might ask, for example, how much a visitor to the Grand Canyon would be willing to pay to reduce air pollution and improve visibility and scenic views in the park. The results of these surveys can then be used to estimate the value of a scenic view, the protection of biodiversity, or other nonmarket goods. In the water example with which we began, one might quantify or "monetize" the health benefits of cleaner water by considering what people pay to reduce heavy metals in their water: how much do people pay for water filtration systems, for example? Or using contingent valuation, how much do people say they would be willing to pay for cleaner water?

Hedonic pricing and contingent valuation attempt to address a critical concern about CBA, the charge that it fails to account for nonmarket goods such as health and aesthetic values. It may have already occurred to you that neither of these methods is perfect. For example, what people actually *do* pay as well as what they would be *willing* to pay is, in part, a function of what people are *able* to pay. A person with a substantial income may pay a significant premium for fresh, organic vegetables, whereas another person, whose budget forces a choice between food quantity and food quality, might choose a larger amount of lower quality, less expensive food. This does not necessarily show that the latter individual prefers low-quality food to high-quality food, but only that the decision is constrained by other variables. Similarly, what the Gates Foundation is willing to pay to reduce the incidence of severe diseases in Africa greatly exceeds what I am willing to pay, but this in good part reflects the fact that the Gates Foundation's resources vastly exceed my own. There are other worries about these modes of valuation, as well, some of which will be discussed later.

Proponents of CBA see this as a critical way to make systematic decisions about proposed environmental regulations or policies, taking into account both upsides and downsides. Advocates of CBA have pointed out that legislators, regulators, and the public can sometimes be afflicted by tunnel vision: a nuclear accident, for example, may heighten the salience of radiation risks and lead to stringent new regulations on the approval of new nuclear power plants. If we focus only on radiation risk reduction in establishing such regulations, we may fail to balance the benefits of the regulation against other important costs, such as increases in the price of power, or increased demand for other forms of power, such as coal, that damage human health and the environment in other ways. CBA, it is argued, helps bring *all* the costs and benefits into view.

Critiques and alternatives: Objections to utilitarianism and cost–benefit analysis

We have considered utilitarianism, CBA, and their similarities. Although the discussion has not been entirely uncritical, I have tried to highlight the appeal of both approaches. It is now time to consider some concerns. The first set of concerns has to do with utilitarianism specifically, but many (though not all) of these critiques carry over to CBA. Recall that utilitarianism is a consequentialist moral theory: consequences are *all* that matter morally for utilitarians. Thus, as suggested earlier, *nothing is ruled out in principle*: lying, torture, and stealing are not categorically forbidden, for example. In practice, it may be rare that utilitarianism would support the use of torture. However, torture could be morally required by utilitarianism in some cases, specifically those in which torturing someone—even an innocent—would maximize overall utility. This feature of utilitarianism is one that conflicts with moral intuitions regarding the protection of certain basic rights, and more specifically, the impermissibility of sacrificing an innocent for the benefit of the whole. It also calls attention to the fact that consequences may not be all that matter in ethics. We make commonsense as well as legal distinctions between murder and manslaughter, for example, and these distinctions are based on the intentions behind each act, not their consequences.

A second worry is that *utilitarianism does not take distribution into account.* Utility may be maximized in some cases by producing additional pleasure (or preference satisfaction) for those who are already well off, while creating additional suffering for those who are already disadvantaged. Although such circumstances may be rare, because utilitarianism focuses on overall consequences, the theory does not rule out distributional inequities.

Third, there are *epistemic problems* with utilitarianism. We have already discussed the challenges of measuring—or even more difficult, predicting—the effects of a particular action on utility. These challenges are exacerbated once we ask ourselves which kinds of effects we are to consider (direct effects only, or indirect effects as well?) and how far into the future we should go.

Epistemic problems tie closely to the problem of *commensurability.* Can we really compare all pleasures and pains, or all costs and benefits, on a single scale? Can and should we measure in common terms the value of friendship and the value of inexpensive consumer goods? Some philosophers argue that there are many disparate goods and these simply cannot be aggregated on a single scale. Instead, different goods need to be considered in the course of a deliberative process that avoids reduction of these complex elements into a single bottom line (see, e.g., O'Neill et al. 2008, p. 48).

The problem of commensurability is also tied to the point that *not all preferences may deserve equal consideration.* Utilitarianism's egalitarian view that all preferences count supports preference neutrality in CBA. However, as Mark Sagoff (2012) has pointed out, preferences are a motley crew. Some preferences—like a preference that others suffer—are ethically dubious; in addition, all of us possess preferences of different *kinds,* which in turn warrant different sorts of consideration. For example, we may have preferences as *consumers* that differ from our preferences as *citizens* (Sagoff 2012). As a consumer, I want the most for my money: I will seek the cheapest gas, for example, and will take all the tax deductions for which I am eligible. Nevertheless, as a citizen I might vote for policies that increase the cost of gas, or that eliminate tax deductions that favor the well off at the expense of the poor, even if neither of these changes would benefit me personally. Sagoff's point is that we should favor policy-making processes that take into account the distinction between citizen and consumer preferences. Utilitarianism and CBA collapse these distinctions by remaining neutral about preferences, and by aggregating them.

All of the above-stated objections apply both to utilitarianism and to CBA; however, there are additional concerns that apply only to one or the other. For example, concerns about contingent valuation and hedonic pricing are specific to forms of CBA that employ these tools, and the "special relations objection"—which holds that utilitarian's neutrality is problematic because it prohibits individuals from giving extra weight to outcomes for those, like spouses or children, with whom they have special connections—seems more apt for utilitarianism as a moral theory governing the behavior of individuals than for CBA as an institutional and governmental planning tool.

In the arena of environmental policy, many argue that CBA is the most thorough and systematic approach, because in principle it takes *all* costs and benefits into account. To objections regarding unreliability in the pricing of nonmarket goods, defenders argue that some numbers are better than no numbers: it is better to account for goods such as clean air and water as best we can than to leave them out of the equation entirely. CBA also helps avoid myopic focus on a single risk, at the expense of ignoring others. And, many ask, what are the alternatives? To adapt a well-known saying often used to characterize democracy, CBA is the worst strategy for decision-making—except for all the rest.

But is this true? CBA is a powerful tool for examining alternative policies in cases where effects are diverse and influence large numbers of people. Yet, opponents have raised some serious concerns. Some have proposed employing CBA in a limited way: we might, for example, set a particular social or environmental goal, then employ CBA only to determine the most efficient way to reach that goal. Or we might embrace "CBA with constraints," permitting cost–benefit tradeoffs only to a certain point. In such a scheme, we might stipulate a certain guaranteed base level of water quality, for example, that cannot be traded off against anything else, or a certain base level of safety in the workplace that cannot be sacrificed, no matter how great the collateral benefits.

Other alternatives are more radical, rejecting the basic assumption that costs and benefits are symmetrical and can be quantified and traded against one another. The *precautionary principle*, or what is sometimes known more generally as a precautionary approach, places special weight on the protection of human health and the environment. Under this approach, human health and the environment are prioritized. The precautionary principle suggests that we take measures to protect human and environmental health when our actions place them at risk, even if the precise nature and magnitude of the risks are unknown. Although there are many formulations of the precautionary principle, one prominent version states as follows:

> When an activity raises threats of harm to human health or the environment, precautionary measures should be taken even if some cause and effect relationships are not established scientifically. In this context, the proponent of the activity, rather than the public, should bear the burden of proof. (Science and Environmental Health Network 1998)

Versions of the precautionary principle have been incorporated into a wide range of environmental policy statements, including the Rio Declaration on Environment and Development, the United Nations Framework Convention on Climate Change, the Cartagena Protocol on Biosafety, and the 1992 Treaty on European Union (Wiener 2002). Proponents of precaution defend it as a commonsense approach, embodying the spirit of the old adage "better safe than sorry." Opponents see the precautionary principle as incoherent or irrational, arguing that the principle focuses too much on the risks of new technologies and thereby deprives society of potential gains associated with new drugs, plants genetically modified to deliver more nutrition, or other innovations. We will return to the precautionary principle in Chapter 6, where we discuss responses to global climate change. For now, it is worth considering how the precautionary principle might diverge from CBA in matters of environmental policy, and whether a precautionary approach can serve as a viable alternative.

As we shall see, there are also theories in the Western moral tradition that reject the basic premises of utilitarianism (and that would reject CBA, accordingly). Those concerned that utilitarianism fails to protect rights and allows the sacrifice of individuals to achieve greater good overall may feel the pull of Kant's ethics, with its foundational commitment to respecting all persons.

For further thought

1. What is the greatest strength of utilitarianism, in your view? What is the theory's greatest weakness?
2. Are monetary costs and benefits a good measure for preference satisfaction?
3. If the local utility company proposed to build a new power plant in your town, what kinds of costs and benefits would you recommend it consider? Are some of these costs and benefits easier to quantify than others?

Kantian ethics and the environment

Kant's ethics: Rationality, reciprocity, and respect

The philosophy of Immanuel Kant (1724–1804), a German philosopher of the Enlightenment, has had an enormous impact on the way we understand the relationship between the human mind and the world. Kant proposed that human experience is conditioned by certain frameworks. Thus, although we have access to reality, we can never directly know "things-in-themselves," or the world as independent of our perception of it. Although not everyone accepts Kant's view, his arguments have shaped important ongoing debates in metaphysics and epistemology. For our purposes, however, it is Kant's ethical theory that is of greatest interest. Like many philosophers in the Western tradition, Kant saw rationality as central to our humanity. The key ideas in Kant's ethics are rationality, autonomy, respect, and reciprocity, and these ideas are importantly linked.

It is common to think that there is something distinct about human beings that separates us from other members of the animal kingdom, and that our distinctive humanity provides the ground for the obligations we have to one another. Witness Article 1 of the Universal Declaration of Human Rights (available on the United Nations web site, at http://www.un.org/en/documents/udhr/):

> All human beings are born free and equal in dignity and rights. They are endowed with reason and conscience and should act towards one another in a spirit of brotherhood.

According to Kant, what makes humans special is that we are rational beings. In virtue of our rationality, we are capable of acting morally and also deserve moral consideration. To put it another way, rationality marks out both the class of *moral agents* and the class of *moral patients*. Moral agents are those who are capable of acting morally: plants and insects, for example, are not moral agents and cannot be held morally responsible. If a bee stings you, it is appropriate to be upset or angry, but not *morally* outraged. However, if another person hurts you, then he or she can be morally blameworthy for their act. Those who have moral obligations and whose actions are subject to moral praise and blame are moral agents.

On almost every standard moral view, all moral agents are also moral patients, those to whom we have moral duties. But some moral patients are not moral agents: infants, for example, are not moral agents in any robust sense, yet on almost every standard moral view, we have moral obligations to babies. It seems that Kant's view needs some way of accommodating this intuition, and though Kant sometimes seems to suggest that one must be a moral agent in order to be a moral patient, defenders of Kant such as Allen Wood (1998) and Holly Wilson (2012) have argued that his view is broader than this, encompassing not only actual rational agents but also those who are partially rational or potentially rational, or were previously rational. Wilson (2012, p. 67) holds that any member of the human species "has the potential for rationality even if she never exhibits it." As we shall see, though, Kant's view excludes nonhuman animals from direct moral consideration, and many environmental ethicists have criticized it on this ground.

Before returning to Kant's views on animals, however, we should consider more carefully why he believes that rationality is so crucial. According to Kant's view, humans possess both an animal and a rational nature. Our animal nature is the source of our desires: these push and pull us in various directions, prompting us to seek food, drink, rest, or excitement. If we were driven only by these desires, as animals are, we would, in a sense, be enslaved by them: rather than controlling ourselves, we would be controlled by these unruly forces of desire. Or as Kant puts it, rather than being autonomous (self-ruled), we would be heteronomous (ruled by something other than ourselves). Fortunately, we possess reason, which enables us to act on rational principles rather than on mere inclination, or desire. As self-legislating beings, we can set rational principles of action for ourselves, and we can recognize that other rational beings share this same capacity for self-legislation.

We can see at this point why rationality serves as the lynchpin of Kantian morality. Rationality is what sets us free: it enables us to break the chains of our animal natures and subject our desires to rational control. As such, rationality is a fundamental capacity that must be protected and preserved. Following Kantian reasoning, we can see the fundamental importance of rationality in our own lives, and this insight prompts us each to see our own capacity for autonomy as something inviolable and deserving of respect. Yet, we can also see at this point that autonomy plays an important role not only in our own lives, but in the life of every rational being. *Thus, the autonomy of every rational being is deserving of respect.* This is the core insight of Kantian morality.

We may then ask how we are to respect the autonomy of all rational beings, including ourselves. And to this, Kant says that we should act on principles—or *maxims*—that all could rationally accept. This requires that in all our actions, we treat others as rational beings: we must act in ways that acknowledge others as ends-in-themselves, never as mere instruments to be manipulated for our ends. Kant believes that there exists just one moral principle, which he calls the *Categorical Imperative.* This imperative requires that we respect the rational nature of others. In the *Foundations of the Metaphysics of Morals*, Kant (1995) offers four formulations of the Categorical Imperative, arguing that all four are equivalent. It will suffice to focus on two of these here:

1. The first, known as the "Universal Law Formula," is a principle of rational reciprocity, commanding each of us to act on principles that all persons could rationally accept: "Act only according to that maxim by which you can at the same time will that it should become a universal law" (Kant 1995, p. 38).
2. The second is the "Humanity Formula," which emphasizes our duty to respect others, and to act only in ways that reflect our recognition of others as rational, autonomous beings: "Act so that you treat humanity . . . always as an end and never as a means only" (Kant 1995, p. 46).

The first formula helps highlight what is wrong with cheating, lying, or offering false promises. To act on a principle that permitted these activities would be rationally inconsistent, according to Kant. Imagine that I borrow five dollars and promise to pay you back next week. But later I decide that I'd rather keep the five dollars for myself. The maxim of my action seems to be something like this: "I'll break my promises whenever it is advantageous to me to do so." Yet, if this maxim were universalized, the whole institution of promising would break down: a promise would cease to have meaning because the so-called promises would no longer be reliable, since they would always be contingent on the whims of the promiser. Kant argues that maxims that permit cheating, lying, or false promises lead to rational contradiction, and so fail the universalizability test laid out by the Universal Law Formula. An intuitive way to think about this version of the Categorical Imperative is that it prohibits action on maxims that allow you to make an exception of yourself. In this sense, it is a principle of rational reciprocity.

Kant's Humanity Formula, though arguably equivalent in its implications, highlights another aspect of the Categorical Imperative, focusing our attention on respect for others' autonomy. The Humanity Formula instructs us always to treat others as ends-in-themselves, never merely as means. What does this mean? First, it does not rule out the possibility of using others as means to our ends. Instead, it rules out our using them as *mere* means. It is fine to use your music teacher or track coach as a way to learn to play the piano or run a faster mile. At the same time, the Categorical Imperative requires that you respect the autonomy and dignity of your teachers. We rely on one another all the time to help accomplish our ends, and this is permissible; it is when we treat others simply as tools or machines that we fail in our duties.

The Humanity Formula prohibits not only the most blatant failures to respect others as rational persons, such as slavery, but also manipulation and deception, which also undermine the autonomy of other human beings. If a mining company, for example, knowingly exposes its workers to hazardous conditions and substances without informing those employees of the risks they face, the employers of these workers have failed in their moral duties. Respecting the autonomy of others requires that we avoid withholding critical information from them; more affirmatively, it requires that we *enable* others to exercise their autonomy, to realize their potential as self-legislating beings.

Kantian ethics has strengths that particularly stand out in relation to utilitarianism. First, Kant's ethics captures the fundamental moral intuition that some things simply ought not be done, regardless of their effects on utility overall. While it might promote overall utility if we were to harvest the organs of a healthy person to save five others in need of transplants, almost everyone thinks it would be wrong to do so (of course, people could be wrong about this, but the burden of proof is on those who seek to overturn this strongly held intuition, which coheres with many of our other moral beliefs). In "The ones who walk away from Omelas," Ursula LeGuin (2004) describes a society in which life is peaceful and everyone is perfectly happy—yet this utopian existence is contingent on the imprisonment of an innocent, small child, who is kept locked away in a dark, dank room in utter isolation. LeGuin (2004, p. 282) explains:

> If the child were brought up into the sunlight out of that vile place, if it were cleaned and fed and comforted, that would be a good thing indeed; but if it were done, in that day and hour all the prosperity and beauty and delight of Omelas would wither and be destroyed. Those are the terms.

The narrator observes that most people come to accept this tradeoff, and find it worth sacrificing the welfare of one child for the good of the society as a whole. But a few cannot tolerate it, and it is these few who walk away from Omelas, turning their back on all the happiness of their society, finding the exchange of the happiness of many for the utter deprivation of another too much to bear. The intuition that Omelas' Faustian bargain is morally unconscionable is effectively captured by the ethics of Kant.

In addition to categorically prohibiting certain kinds of acts, Kantian ethics provides a basis for human rights. A commitment to the fundamental dignity and autonomy of each individual supports encoding in international law a set of key protections for human beings. Although there are other potential philosophical foundations for human rights, the autonomy ground historically has been central.

A final strength of Kantian ethics also connects to our earlier discussion of utilitarianism. Although Kant recognized that individuals have diverse ends, and his ethics allows for various paths in life, Kantian ethics does not treat all preferences as equal or as deserving of moral weight. Preferences that violate the Categorical Imperative are not to be honored; they are desires we simply have no obligation to promote or help to fulfill.

Kant's ethics thus diverges from utilitarianism in many ways. Whereas utilitarianism is consequentialist, Kant's ethics is *deontological*. Deontological moral theories focus on moral duties, rather than on outcomes. For Kant, it is not the consequences of one's actions that are critical; instead, an attitude of respect is key (Dillon 2010). *Motives* matter. For Kant, the only thing that can be called good without qualification is a good will. An action that issues forth from a good will—out of respect for the moral law—is a moral act, regardless of the consequences that ensue.

Limitations of Kantian ethics

As you may have surmised, the same elements that some philosophers label as strengths, others see as weaknesses of Kant's ethics. Rather than viewing categorical prohibitions—such as the prohibition on lying—as a positive feature, some argue that Kantian ethics is too strict. When lying helps avert serious, negative consequences, for example, perhaps it is the *right* thing to do. In addition, Kant's moral theory encounters difficulties in cases in which all available courses of action seem to violate the Categorical Imperative, where duties conflict. In facing a moral dilemma where we will inevitably violate the autonomy of one person or another, how do we choose? Although contemporary scholars have worked to answer such questions, Kant's ethics provides limited resources for handling moral dilemmas—and this is precisely where utilitarianism asserts its strength. Also, it is sometimes difficult to know what Kantian ethics requires. Many, for example, have criticized the universalizability test for problems it engenders: How does one formulate the maxim of one's action? There seem to be multiple possibilities in almost every case. How does one discern whether universalization results in contradiction? At times this is not obvious. Finally, Kant distinguishes between two kinds of duties: perfect duties, which are absolute, and imperfect duties, which are duties that admit of degrees. One has an imperfect duty to help others, for example. However, the extent of these imperfect duties is unclear, and we are led to a problem similar to the one described above with respect to conflicting duties: Kant's ethics sometimes provides insufficient guidance regarding what we are to do. In this latter case, of course, one might argue in Kant's defense that there are certain realms—and the realm of imperfect duties is a case in point—in which morality *should* be indeterminate, leaving it up to each individual the extent to which he or she helps the poor, develops his or her talents, and the like.

The general shape of Kant's ethics, as well as its strengths and limitations, provides important philosophical background for understanding the relationship of this theory to environmental ethics. Like most classical moral theories, Kant's ethics makes no mention of obligations to the natural world, and animal and environmental ethicists have criticized Kant's explicitly stated views regarding animals. Because of the emphasis Kant places on rationality, and the distinction he draws between humans as rational creatures and other living things as heteronomous, or dictated by desires, Kant's ethics gives animals distinctly lower status than humans. Standard interpretations of Kant, in fact, suggest that animals are excluded from direct moral consideration altogether: any moral duties we have in relation to animals are indirect duties *regarding* animals not direct duties *to* animals. For example, although we have duties *regarding* our pets, these duties are ultimately grounded in the Categorical Imperative, and our duties to other rational beings, rather than in any direct obligation to animals themselves.

Extending respect

Given Kant's views on animals and his deep anthropocentrism, one might think that the theory has little to offer to environmental ethics. However, the core idea of respect has been adapted to enable extensions to animals and the environment more broadly. As we shall see in Chapter 3, the works of both Tom Regan, an advocate of animal rights, and Paul Taylor, who defends a biocentric environmental ethic, draw on concepts of respect and inherent value that have strong roots in the Kantian tradition. Where both authors depart from Kant is in the scope of moral considerability.

In simplest terms, an entity is morally considerable if and only if it "counts" from a moral perspective: if it can be morally harmed, and if is the sort of thing to which we can have duties or obligations. Although in some sense I can damage or "harm" a piece of paper by tearing it up, the paper cannot be *morally* harmed by my tearing or crumpling it. If a boy tears up a piece of paper bearing the artwork of his sister, then he has wronged his sister, but not the paper itself. We may have duties to respect others' things, but in many cases these are duties to the owner of those things, not duties to the things themselves.

For Kant, only rational beings are morally considerable. Rationality is the only thing that has inherent value, and to which we have moral duties. For Regan, by contrast, all living, experiencing beings have inherent worth, and this inherent worth grounds moral rights: "All [things with inherent value] have an equal right to be treated with respect, to be treated in ways that do not reduce them to the status of things, as if they exist as resources for others" (Regan 2012, p. 86). Regan here clearly echoes the key idea of Kant's Humanity Formula, emphasizing the idea that all "experiencing subjects of a life" deserve to be treated as ends-in-themselves, rather than mere resources for our use and disposal.

Similarly, Paul Taylor retains the notion of respect but expands the circle of moral considerability further to include all living things. For Tom Regan, plants lack subjective experience and hence have no inherent worth, whereas Taylor argues that "every organism, species population, and community of life has a good of its own which moral agents can intentionally further or damage by their actions" (Taylor 1981, p. 199). Taylor believes that there exist good reasons for adopting an ultimate moral attitude of respect for the natural world, and part and parcel of that attitude is recognition of the intrinsic value of all living things.

Taylor and Regan share a commitment to the following idea: all things that possess inherent value possess it equally. As for Kant, respect and intrinsic value are conceptually linked, and we are not licensed to sacrifice respect in the case of a few for the benefit of the many. The challenge with such a view, as noted above, is to reconcile competing claims. Nevertheless, the idea that a moral attitude of respect is critical to an adequate environmental ethic is one worthy of further consideration. Another way of incorporating respect into an environmental ethic might be to focus not on respect as an absolute duty, but on respectfulness as a virtue, which leads us to virtue ethics, the third major moral theory we will discuss.

For further thought

1. Explain how Kant's Universal Law Formula and his Humanity Formula reflect different statements of the same fundamental moral principle.
2. How might you argue that an extended Kantian ethics offers a stronger basis for an environmental ethic than utilitarianism? How might you argue the reverse?

Virtue ethics and the environment

Aristotle's teleological ethics

While they differ in many respects, Kantian ethics and utilitarianism have in common a focus on what makes *acts* right or wrong. Virtue ethics, by contrast, asks different questions: "What makes someone a good person?"

or "What is it to live a good human life?" These questions characterize the ethics of ancient Greece, and also resonate strongly in classical Chinese traditions such as early Confucianism. Although virtue ethics can be traced to many sources, the *locus classicus* is Aristotle's *Nicomachean Ethics* (1999). Interestingly, whereas modern ethical theories often draw a contrast between what is good for oneself, on the one hand, and what is morally good, on the other, Aristotle's ethics brings these together. According to Aristotle's view, to live a good human life—a rich, fulfilling, and meaningful life—is to live a life of moral virtue. Another way of putting this is that "the virtues benefit their possessor" (see Hursthouse 2001, ch. 8, for discussion), a view that has roots in the thought of Socrates and Plato.

Aristotle's optimistic conviction on this score ties closely to his teleological worldview. Aristotle envisioned animals, plants, and human beings as part of a larger organic order in which each kind of living thing has a particular purpose, or *telos*, for which it is uniquely suited. This *telos* serves as an ultimate end: to realize one's *telos* is to live well: to be *eudaimon*, happy or flourishing. Thus, what makes an organism a *good* organism of its kind is to realize its *telos*. It is for this reason that Aristotle's ethics is called *teleological*: it starts by establishing the good, or end, that is to be promoted. The good in Aristotle's ethics is flourishing or *eudaimonia*, just as the good in utilitarianism is the maximization of utility, and *Nicomachean Ethics* is fundamentally an explanation of how we can achieve that good.

For Aristotle, the notion of *telos* is closely related to the idea of function. Both organisms and artifacts (intentionally made human products) have functions in his view, and a good organism or artifact of its kind is an organism or artifact that performs its function well. If the function of a knife is to cut, then a good knife is a knife that cuts well. Aristotle argues that the function of human beings is derived from our unique capacities for reason. Thus, a good human being is a human being that reasons well.

It is important to note that Aristotle's conception of reason is significantly broader than one might think. By saying that exercising our rationality is essential to living well, Aristotle did not mean to suggest that to flourish is merely to excel in one's logic class, or to ace calculus. The human *telos* involves reasoning well about what to believe *and* what to do. For Aristotle, reason included both the theoretical reasoning common to math, philosophy, and the sciences as well as *practical reason* (or practical wisdom). Whereas theoretical wisdom concerns universal and timeless truths, practical wisdom concerns things involving human action, about which we can decide:

> [Practical wisdom] . . . is about human concerns, about things open to deliberation. For . . . no one deliberates about things that cannot be

> otherwise, or about things lacking any goal that is a good achievable in action. (Aristotle 1999, 1141b10–1141b14)

The extent to which Aristotle identified human flourishing with the exercise of theoretical reason is a matter of some scholarly dispute; however, most of *Nicomachean Ethics* is dedicated to explaining what is required to live a life guided by practical reason, or *phronesis*, and it is on this that we shall focus.

As noted above, a good human life, according to Aristotle, is a life of virtue, and a truly and fully virtuous person is practically wise. But how are we to achieve such a life? Rather than focusing on a criterion of right action or a delineation of categorical duties, Aristotle explains how we develop the virtues, outlining a path of self-cultivation. Moral virtue, according to Aristotle, begins with habit. How does one learn to be generous? By doing generous acts. How does one learn to be courageous? By doing courageous things. A virtuous moral character is produced through and reflected in habitual action. The development of virtue begins at a young age, as our parents encourage us to develop habits consistent with the virtues. In developing the virtues, we can pattern our behavior after those paradigmatic individuals who most fully reflect a virtuous character, individuals who possess practical wisdom.

Yet emulating the *phronimos*, or practically wise person, is not sufficient to become fully virtuous oneself, because a fully virtuous person must be able to assess in a nuanced and sensitive way what each situation requires. This, in turn, requires moral experience. We learn through experience that courage, for example, does not always require a display of personal valor on the battlefield. Courage can also involve quietly but insistently standing up against racism or sexism, or questioning dominant ways of thinking such as the view that success is best measured by material wealth. To know what courage, or any other virtue, requires in diverse contexts requires moral perception (see Blum 1991; McDowell 1998; Jacobson 2005; Lear 2013). This perceptual ability enables the virtuous person to detect morally relevant aspects of a situation that others might miss: moral perception makes certain details salient (Blum 1991). What's more, moral perception allows us to see a situation in an integrated way: Aristotle thought that the fully virtuous, or practically wise, person could assess a situation not only through the lens of a single virtue, but through an integrated understanding of how all the relevant virtues fit together. A fully virtuous person can see what combination of compassion and justice is required in a given instance, for example.

Aristotle believed that the fully virtuous, or practically wise, person possessed the ability to integrate the virtues, and this is tied to his *unity of the virtues* doctrine. According to this doctrine, anyone who fully possesses one virtue must possess them all. As explained at the end of book VI of

Nicomachean Ethics, to be genuinely and fully good requires practical wisdom, and practical wisdom requires possession of all of the virtues:

> What we have said, then, makes it clear that we cannot be fully good without [practical wisdom], or [practically wise] without virtue of character. (Aristotle 1999, 1144b30)

Here Aristotle distinguishes between *natural virtue*, which Rosalind Hursthouse (2012) describes as "a proto version of full virtue" that can be possessed by children or others who lack practical wisdom, and *full virtue*, which is virtue guided by practical wisdom. Since one must possess practical wisdom to possess any of the virtues in its fullest sense, and since the possession of practical wisdom entails the ability to integrate all of the virtues in one's decisions, a person who possesses practical wisdom must possess all of the virtues: one cannot be fully just without understanding how justice and compassion fit together in one situation, or how generosity and temperance come together in another. Just as we call a musician a virtuoso only if she plays in a way that brings together mastery of all the critical elements of her instrument, so a *phronimos*—a moral virtuoso—is one who possesses all the virtues and lacks in none.

Strengths, weaknesses, and contemporary developments in virtue ethics

As we have seen, Aristotle's ethics has a stronger emphasis on the diachronic dimension of moral life than either utilitarianism or Kantian ethics. Although Aristotle believes that the virtuous person is a person who acts well in any given instance, *Nicomachean Ethics* concerns itself more strongly with the character and patterns of behavior of the virtuous person than with a criterion of right action. This has been viewed by some contemporary moral philosophers as a weakness, and by others as a strength. Those who see it as a weakness charge that unless one is virtuous (and no one starts out this way), one cannot know the right thing to do, according to Aristotle's theory. Instead, one must rely on imitating moral models (knowing which models to emulate) and on habits inculcated by teachers and parents. On the other hand, while it may be quite straightforward in theory to determine what is right from a utilitarian point of view, it is much more complex in practice, and we have already seen that Kantian ethics runs into difficulties in clearly establishing in each case what one ought to do. So, from a comparative perspective, perhaps virtue ethics is not so bad off after all.

What's more, as noted above, there are those who see the de-emphasis on individual acts as a strength of virtue ethics: for what is the good of knowing clearly the right thing to do if one lacks the wherewithal to do it? With its emphasis on moral development and the cultivation of appropriate patterns of behavior, virtue ethics is thought by many to offer a more psychologically realistic moral theory and one that provides richer guidance in living a moral life. Virtue ethics draws attention away from the puzzling but unusual moral dilemmas—such as the famous "trolley problem" (Thomson 1985; Foot 1978)—that preoccupied many twentieth century analytic moral philosophers and focuses more strongly on the mundane, daily challenge of living a good human life and acting well in relation to others.[2]

Aristotle's ethics in its original form, of course, incorporates ideas that many contemporary philosophers find problematic. For one, as noted above, Aristotle's ethics relies on the notion that each kind of being has its own distinctive function, and many have questioned the metaphysical basis for this claim. Is there really a single, unitary function for human beings, *as* human beings? Furthermore, Aristotle's teleological worldview is hierarchically organized, with organisms lower in the hierarchy existing for the sake of those higher up. This, too, seems questionable—and since humans are near the top of the hierarchy, it is a strongly anthropocentric view, inimical to theories of environmental ethics that seek a broader and more egalitarian sphere of moral concern. Finally, some have questioned the adequacy of the specific virtues that figure in Aristotle's ethics. While few would contest the importance of courage or generosity, some of Aristotle's virtues—such as magnificence, having to do with "expenditure that is fitting in its large scale" (Aristotle 1999, book IV)—have an aristocratic flavor, and there are other virtues—such as humility or respect—that he seems to overlook.

Despite these weaknesses, many contemporary moral philosophers have developed renewed interests in virtue ethics, in part because it addresses aspects of our moral lives that ethical theorizing in the twentieth century largely ignored. Virtue ethics is more attentive to aspects of moral psychology and moral development than the main alternative theories, for example. What's more, many contemporary virtue ethicists believe that virtue ethics can survive without the hierarchical and teleological metaphysics of Aristotle, and that the specific list of virtues outlined by Aristotle is not fixed. Virtue ethics has thus expanded in many new directions, and even committed Kantians and consequentialists have taken an interest in how virtue may play a role. Philosopher Julia Driver (2004), for example, defends a theory of "virtue consequentialism," and Kantian scholar Thomas E. Hill, Jr has both explored virtue in Kant's moral philosophy and written a groundbreaking essay in environmental virtue ethics (see Hill 1983 and 2012). The exploration of virtue ethics is, in fact, an important development in recent environmental ethical theory, and the following section describes the relationship between virtue ethics and the environment in greater detail.

Environmental virtue ethics

On reading *Nicomachean Ethics*, one might quite reasonably conclude that Aristotle offers little to environmental ethics. Like the other classical moral philosophers whom we have discussed, Aristotle has virtually nothing to say about our ethical relationship to the environment *per se*. The realm of ethical relations for Aristotle is the human realm: it is in our interactions with one another that we express or fail to express moral virtue. Nevertheless, the concepts of character and virtue that figure in Aristotelian virtue ethics are concepts we frequently encounter in discussions of our relationship to the natural world. Environmental problems such as pollution, deforestation, and climate change, for example, are often described as the result of vices such as greed, carelessness, and hubris, while those seeking to repair our relationship to the natural world emphasize humility, respect, and care as key virtues. The language of virtue and vice is thus not unfamiliar in the environmental arena.

In 1983, Professor Thomas Hill published an essay that opens with the following anecdote:

> A wealthy eccentric bought a house in a neighborhood I know. The house was surrounded by a beautiful display of grass, plants, and flowers, and it was shaded by a huge old avocado tree. But the grass required cutting, the flowers needed tending, and the man wanted more sun. So he cut the whole lot down and covered the yard with asphalt. After all it was his property and he was not fond of plants. (Hill 1983, p. 211)

Commenting on this story, Hill notes that the relevant ethical question is not so much whether the man had a *right* to pave his yard, or whether he failed to maximize utility by doing so. Instead, Hill argues, a case like this prompts us to ask, "What sort of person would do a thing like that?" With this question, Hill draws our attention to *character*. He is not asking whether it is wrong to pave one's yard, and if so, why. He is not looking for a moral principle that would prohibit such actions. Instead, we are prompted to look at the person, and how actions flow from an individual's character. What kind of person would remove every living thing from his or her yard in order to cover it with asphalt? From what attitudes, dispositions, habits of action, and ways of seeing the world does such behavior proceed? We can then ask, what is problematic—if anything—about these attitudes, dispositions, habits, or ways of seeing the world?

The sort of person who would tear out all his plants and pave his yard with asphalt is one who seems indifferent to nature, and Hill explains:

> [I]ndifference to nonsentient nature typically reveals absence of either aesthetic sensibility or a disposition to cherish what has enriched one's life

> and . . . these, though not themselves moral virtues, are a natural basis for appreciation of the good in others and gratitude. (Hill 1983, p. 216)

The idea highlighted here is that indifference to nature typically reflects a lack of appreciation for beauty or for those things on which the quality of one's life depends. These deficiencies in turn hinder the development of the virtue of gratitude. As Hill suggests, gratitude is a fundamental human excellence, or virtue. Thus if we seek to cultivate gratitude, we should also cultivate a sensitivity to nature, on which a grateful character may in part depend.

Gratitude, of course, is just one environmentally relevant virtue, and a theory of environmental virtue ethics should offer a fuller picture of what the key virtues are. There are various ways of specifying environmental virtues, and Ronald Sandler (2005) lays out four possible strategies. The first, which Sandler calls extensionism, begins from standard interpersonal virtues (such as the Aristotelian virtues), then argues that these virtues extend to nonhuman domains. Thus, "if compassion is the appropriate disposition to have toward the suffering of other human beings and there is no relevant moral difference between human suffering and the suffering of nonhuman animals, then one should be compassionate toward the suffering of nonhuman animals" (Sandler 2005, p. 4). Environmental virtues can also be grounded in relation to the idea that the virtues are traits of character that benefit their possessor. Insofar as humility toward nature, for example, benefits those who possess such humility, or aesthetic sensitivity to natural beauty enriches its possessor's life, these traits count as environmental virtues. Authors such as biologist E. O. Wilson (1984) and journalist Richard Louv (2011), who argue that human contact with nature is critical to our health and well-being, provide support for the idea that it is in our own enlightened self-interest to cultivate certain virtues in relation to the environment. A third approach emphasizes the role of environmental virtues in human excellence, or the achievement of *eudaimonia*. On this view, to live a good human life or to flourish (in something akin to the Aristotelian sense) is not only to possess virtues that make for good social relationships but also to possess virtues that make for appropriate ecological relationships. This approach expands the account of flourishing to encompass the relationship between humans and nature, and with this account in hand, we can see the virtues needed to achieve this broadened conception of flourishing.

Lastly, we might identify key environmental virtues by appeal to paradigmatic environmental role models such as Rachel Carson, Aldo Leopold, or Henry David Thoreau. Rachel Carson was a biologist and writer who worked for the United States Bureau of Fisheries, and later for the U.S. Fish and Wildlife Service. She is the author of the classic book, *Silent Spring*, which publicized the dangers of pesticides to humans and the environment. Aldo Leopold, an

American conservationist educated at Yale School of Forestry, worked for the U.S. Forest Service for many years, and later became a professor at the University of Wisconsin-Madison. Throughout his career, he pushed toward more holistic and ecologically informed practices of land management, and is well known for his posthumously published book, *A Sand County Almanac*, which includes his famous essay, "The Land Ethic." Thoreau, of course, is known for *Walden*, which chronicles his time living in a small cabin in the woods on Walden Pond, not far from Boston. Environmental philosopher Philip Cafaro (2001) argues that Carson, Leopold, and Thoreau exemplify important environmental virtues and serve as important models for our own moral development. Cafaro highlights simplicity as a key virtue for Thoreau, attentiveness to nature as an important trait of Leopold's, and humility in relation to nature as central for Carson.

Virtue ethics in general, and environmental virtue ethics in particular, provides a distinctive moral perspective that may be especially valuable in the environmental realm. Although we often think of ethics as involving a set of requirements and prohibitions—"oughts" and "ought nots"—virtue ethics gives us an aspirational vision of a good person toward which we can work. Rather than conceive our ethical task as one of monitoring our individual actions for rightness, we can instead focus on the long-term cultivation of habits of mind and action that exemplify the virtues. As we have seen, acting well and living well from a virtue ethical point of view involve not only certain behaviors, but certain ways of perceiving the world and responding to it emotionally. The virtuous person does the right thing, at the right time, for the right reason, and is able to do so because he or she is attuned to the world in a way that the vicious, or even the morally indifferent person, is not.

Of course, every theory has its detractors, and critics of virtue ethics worry that it is too inwardly focused, too "agent-centered." As environmental philosopher Holmes Rolston III suggests, what we need to worry about from an ethical perspective is the world, not ourselves:

> But put first things first: life in nonhuman (and human) others, and second things second, one's virtues. Life will be inadequately reverenced if I respect the lives of others with the increase of my virtue in mind. The foundation here is a life ethics, not a virtue ethics. (Rolston 2005, p. 73)

Although scholars will undoubtedly argue over the relative importance of virtue in an environmental—or general—ethical theory, the exploration of virtue has generated some promising new directions for environmental philosophy. In particular, virtue ethics has prompted greater attention to moral development, moral motivation, and the possibility that cultivating virtues in relation to the natural world might be part of a flourishing human life.

For further thought

1. Is Aristotle's view of the human *telos* plausible? Is there a single, unitary good for human beings, invariant across time and space?
2. Of the four grounds for environmental virtue described by Ronald Sandler, which are most promising?
3. What sorts of habitual practices might help cultivate environmental virtues? Are good habits sufficient for the development of environmental virtue?

Conclusion

We have examined in detail three classical Western moral theories and their relevance to environmental ethics. Although each of these theories has limitations—in some cases significant limitations—with respect to animals and the environment, we have seen that even a deeply anthropocentric theory, such as Kant's, may have conceptual resources to offer an expanded conception of our obligations to the natural world. Nevertheless, traditional Western moral theories were not developed with nonhuman organisms or the environment in mind, and at the very least, will require fairly extensive modification or extension in order to take the natural world more fully into account. Many environmental philosophers have argued that the resources of the standard moral theories are inadequate, and that we need to build a new, environmental ethics from the ground up. The anthropocentrism of classical theories such as Kantianism or Aristotelian ethics has been a key sticking point. Thus environmental philosophers have spent significant effort developing nonanthropocentric views and arguing for the expansion of moral consideration to include animals, plants, species, ecosystems, and the natural world as a whole. Chapter 3 takes up the challenge of broadening the circle of moral concern, and the various strategies for doing so.

Before turning to that, however, it is worth saying a word about the role and relevance of moral theories. Although many moral philosophers identify with a particular ethical approach—consequentialism, deontology, or virtue ethics—and work to develop a version of their preferred approach that can withstand critique and answer the most serious objections, moral philosophy need not be a matter of finding and refining a single theory, a search for the Holy Grail of ethics. Instead, we can see theories—as some philosophers of science do—as models. On this view, a single theory can never capture fully all the complexity of the phenomenon it attempts to characterize;

instead, theories highlight critical elements of the phenomenon in question. For complex phenomena like ethics, there may be no single model or theory that addresses all aspects of the phenomenon, but there may be more than one theory that is apt. Fully understanding ethics thus might require the perspectives of more than a single theory. On this view, theories are like lenses that allow us to focus on different dimensions of moral life: utilitarianism directs our attention to consequences; Kant's ethics to respect and autonomy; Aristotle's virtue ethics to character and its connection to living well. If we accept this perspective, then moral reasoning will require a kind of intellectual flexibility that enables us to see ethical issues from multiple points of view. Some philosophers go further and reject normative ethical theories altogether as guides for action. Moral particularists hold that morality cannot be codified, thus "moral principles" are crude efforts to capture complexities that no set of rules can adequately describe:

> Moral Particularism, at its most trenchant, is the claim that there are no defensible moral principles, that moral thought does not consist in the application of moral principles to cases, and that the morally perfect person should not be conceived as the person of principle. (Dancy 2009)

Even if we reject both particularism and the "theories as models" view and ally ourselves with a particular normative ethical theory, it is important to remember that moral theories are not algorithms. The application of normative theories to real-world problems and cases is itself a complex task involving interpretation and supplementary assumptions. Thus, utilitarians may disagree about whether and what kind of animal experimentation is permissible, just as Kantians may disagree about the morality of abortion. Ethical deliberation is hard work, and though theories may shape and aid our deliberation, we should not expect normative ethical theories to singlehandedly answer all of the ethical questions we face.

Further reading

Aristotle (1999). *Nicomachean Ethics* (2nd edn). Translated by Terence Irwin. Indianapolis, IN: Hackett.

Cafaro, P. (2001). "Thoreau, Leopold, and Carson: Toward an environmental virtue ethics." *Environmental Ethics*, 22, 3–17.

Hill Jr., T. E. (1983). "Ideals of human excellence and preserving natural environments." *Environmental Ethics*, 5, 211–224.

Kant, I. (1995). *Foundations of the Metaphysics of Morals*. Edited by Lewis White Beck. Upper Saddle River, NJ: Prentice-Hall.

Mill, J. S. (1979). *Utilitarianism*. Indianapolis, IN: Hackett.

Regan, T. (1987). *The Case for Animal Rights*. Dordrecht, The Netherlands: Springer.
Sagoff, M. (1990). *The Economy of the Earth*. New York: Cambridge University Press.
Singer, P. (2002). *Animal Liberation*. New York: HarperCollins.
Taylor, P. (1981). "The ethics of respect for nature." *Environmental Ethics*, 3, 197–218.

3

Anthropocentrism and its critics: Broadening moral concern

Introduction: Intrinsic value and moral standing

The three classical moral theories we have discussed—utilitarianism, Kantian ethics, and Aristotelian virtue ethics—have been criticized as overly anthropocentric, or human-centered. Each is arguably anthropocentric in a different way. Utilitarianism, though its traditional form counts the pleasure and pain of all sentient beings, has rarely been applied in a way that takes animal suffering seriously. Furthermore, utilitarianism has no basis for taking account of nonsentient animals, plants, or ecological systems. Kant's ethics is even more anthropocentric—or at least, "ratiocentric"—than that of the utilitarians, for Kant holds that only rational beings have moral worth. Aristotle's ethics also places humans at the center of the moral universe, drawing on a teleological worldview in which "plants exist on account of animals . . . and other animals for the sake of man" (quoted in Gruen et al. 2013, p. 15). While Aristotle acknowledges the dependence of humans on animals and the natural environment for food and other resources, his account of moral virtue and human flourishing does not explicitly consider human relations to animals, plants, or the environment. Each of these canonical theories nevertheless may offer conceptual resources through which to develop a more robust and inclusive environmental ethic, and this chapter offers a more detailed characterization of efforts to broaden the circle of moral concern, both through and independent of traditional Western moral theories.

As the field of environmental philosophy developed in the middle and late twentieth century, many scholars believed that the lynchpin for an adequate environmental ethic lay in clarifying and properly identifying the subjects of moral concern. If anthropocentrism—the view that humans are the only or primary subjects of moral value and moral obligation—was the problem, then nonanthropocentrism seemed the obvious solution. (Of course, the problem could lie not in anthropocentrism per se, but in the particular prevailing *forms* of anthropocentrism, a point to which we shall return.)

Environmental philosophers diagnosed the problem with anthropocentrism as a failure to value nonhuman organisms and other aspects of the natural world. Thus, questions of value have been at the core of discussions of anthropocentrism and nonanthropocentrism. The issue is not that anthropocentrism fails to acknowledge that the nonhuman realm has *any* value. Even anthropocentric theories grant that animals, plants, wetlands, meadows, and the like can have value of *some* kind. But according to anthropocentric ethics, other-than-human things have value only *instrumentally*, as means to the ends of those things that have *intrinsic value*, which is more fundamental. One can think of intrinsic value as value that a thing has in its own right, or for its own sake. It is sometimes referred to as *inherent* value, suggesting that intrinsic value is value contained within the entity that possesses it, not value bestowed from outside. The trouble with valuing organisms and the environment only instrumentally, or extrinsically, according to nonanthropocentrists, is that these modes of valuing are too contingent on human interests and desires. If people do not care about polar bears, or if they do not depend on them for their survival or flourishing, then anthropocentrism provides no ground for a moral obligation to protect polar bears from harm, or the species from extinction. This, many have thought, conflicts with the fundamental moral intuition that there is something wrong with killing animals, razing mountains, or extinguishing species, even if none of our own interests or desires are undermined by doing so.

In his classic article, "Is there a need for a new, an environmental, ethic?" philosopher Richard Routley (2013, pp. 43–44) offers a thought experiment that helps bring this intuition into focus. He describes the following case, which he calls the *last man* example. In this example, there is just one person left on the planet, and this person decides to destroy all remaining forms of life, to the extent he is able. The case brings into focus questions about the nature and grounds of our obligations to other living things. Would it be right to destroy animals and raze forests, if no other humans were around to be bothered by it? Here is Routley's analysis:

> What [the last man] does is quite permissible according to [anthropocentrism], but on environmental grounds what he does is wrong. Moreover

> one does not have to be committed to esoteric values to regard Mr. Last Man as behaving badly. (pp. 43–44)

Routley acknowledges that prevailing moral theories permit destruction of the environment: he notes that "on the prevailing view man is free to deal with nature as he pleases, i.e., his relations with nature, insofar at least as they do not affect others, are not subject to moral censure" (2013, p. 41). However, the common intuition that the last man is acting wrongly reveals a problem with this view. It is here that reflective equilibrium comes into play: should we trust prevailing moral principles and conclude that we are mistaken in our intuitions about the last man case? Or should we trust the intuition that it would be wrong to wantonly destroy the natural world, even if no *person* would be negatively affected by such destruction? On Routley's view, it is the intuition to which we should be faithful: our response to the last man example should call into question our confidence that we are free to deal with nature as we please. Routley rejects the idea that our direct obligations extend only to other human beings, and that any duties regarding the natural world must be derivative from and dependent on these. Instead, he suggests, we can have direct obligations not to harm trees, animals, or other elements of the natural world.

Although he does not put forth his argument in terms of value, Routley indirectly invokes ideas of moral standing and moral considerability, which philosophers often tie to intrinsic value. *Moral standing* and *moral considerability* are synonymous terms that refer to an entity's moral status. As noted in Chapter 2, an entity is morally considerable (or has moral standing) if and only if it "counts" from a moral perspective: if it can be morally damaged, harmed, or disrespected, and if is the sort of thing to which we can have duties or obligations. Although philosophers may quibble about the details, it is typically held that an entity has moral standing if and only if it is a bearer of intrinsic moral value. On this view, if an entity is to count morally, it must be intrinsically valuable, and anything with intrinsic value should be morally taken into account. (Kant's moral theory clearly reflects this structure: a good will—which for Kant is a rational will—is the only thing with intrinsic value and the sole source of our moral duties.)

Lest these debates over intrinsic value and moral considerability seem too abstract, let us turn to a real world case involving related issues. In the early 1970s, the Sierra Club sued the U.S. Forest Service for permitting the development of a ski resort proposed by Disney Corporation in the Mineral King valley, close to California's Sequoia National Park. The club alleged that development of the ski area would damage National Forest and National Park lands, violating environmental laws. However, in *Sierra Club vs. Morton* (1972, 405 U.S. 727) the U.S. Supreme Court found that the Sierra Club lacked standing to sue. More specifically, the Court refused to consider the club's complaints, arguing that it had failed to demonstrate that its members would be directly harmed by the proposed development. To have legal standing to sue in the United States, one must show that one has been or will be injured by the actions of the defendant. Typically, injury is clearest where the defendant's actions have damaged or threaten to damage the plaintiff's *economic* interests; however, other kinds of injuries can be taken into account, as the court noted in its opinion on this case. What is critical, however, is that the parties issuing the complaint be harmed by the defendant's actions.

This legal standing requirement seems to reinforce Routley's point that harm to the environment is considered permissible as long as it does not harm other *people*. Although U.S. law does establish certain basic regulations designed to protect the environment, citizen enforcement of these regulations requires the identification of individual people who have been or will be harmed by their breach. The U.S. doctrine on standing aims to reduce the incidence of frivolous lawsuits, but in doing so leaves the environment vulnerable. In response to the Mineral King case, legal scholar Christopher Stone (2010) argued that the law should be modified to grant standing for natural objects. That way, a group like the Sierra Club could sue on behalf of the forest itself, rather than on behalf of club members whose interests would be undermined by damage to the forest.

In Stone's view, the lack of standing for rivers, forests, and animals is one element of the larger problem that natural entities lack legal rights. Thus, Stone not only challenges a doctrine of standing that prohibits legal action from being initiated to protect natural objects in the absence of injury to human beings, he also argues for a broadening of legal rights so that legal remedies take account of harm to natural objects and respond by repairing these injuries. Under the legal system Stone criticizes, if a landowner downstream from a manufacturing plant sues the company for polluting her water, the legal remedy may be to compensate the downstream landowner financially, or to provide an alternate water source. However, this may leave the river contaminated, because the river itself is not legally considerable.

There are clear parallels between moral and legal considerability: where moral considerability involves whether an entity counts morally, legal considerability involves whether an entity counts legally. Although the two kinds of considerability need not run in parallel (not every moral requirement is codified in the law, for example—it may be wrong to lie to your parents, but it is typically not illegal), they are importantly related. Legal systems generally reflect the central values of a society, and laws can also reinforce or generate values. As legal scholar Jedediah Purdy (2013, pp. 931–932) explains:

> Environmental law is one of the settings where ethical development takes place. This happens not just in law's internal processes, such as standing, or in the pronouncements of courts. At least as important is environmental law's shaping and framing of experience. In experience, new kinds of ethical claims become available, even obvious, which would once have seemed strange.

Giving trees legal standing would open up new ways of understanding their value. Implementing Stone's proposal would require us to think differently about compensation, for example, forcing us to consider questions such as, "How can this forest be repaired? What would be best for the forest ecosystem?" Such questions encourage us to think from the perspective of other living things or from an ecosystem perspective. Though some may object to the idea that an ecosystem can have a "perspective" or can be benefited or harmed, Stone argues that we think and talk about the good of all sorts of other things—like corporations, financial markets, or educational institutions—all the time. So why not trees, rivers, or ecosystems?

The point about reciprocal influence between law and ethics applies more generally, however, and remains relevant even if one is reluctant to accept Stone's claims about rivers and trees. Even something as simple as a mandatory recycling program may change the way people understand their consumption. Recycling, as small a step as it may be, enhances our awareness of what is *trash* as opposed to what is recyclable. This may make us more attentive to the way things are made and to question the role of "disposables" in our lives. At the very least, new legal structures create the *possibility* of new modes of experience and new ethical perspectives.

Philosopher Anthony Weston makes a complementary point in his book, *The Incompleat Eco-Philosopher* (2009), in which he argues that ethical disregard for animals and the environment is facilitated by their degradation in practice. When we house animals in dirty, crowded, repulsive conditions, they appear to us as dirty, degraded beings, not *worthy* of respect. Weston (2009, ch. 3) calls this "self-validating reduction": the diminishment of animals kept in factory farms or the despoilment of an urban streamside by trash makes it harder for us to see anything of value there. We can break this cycle, he suggests, by reversing it. If we treat animals well and see them flourishing, we will find it harder to justify their degradation. Theory and practice—or moral values and action—are intimately linked, each informing the other. By changing our practices, we can change our values, and by changing our values, we can change our practices. It is this latter strategy that environmental philosophy has traditionally pursued, seeking to extend moral consideration to animals, living things, and ecological systems as a whole.

For further thought

1. Do you agree that an anthropocentric worldview would permit the Last Man to destroy nature if there were no other humans left alive? Why or why not?
2. How are moral and legal considerability related? Give an example of a situation in which changing one might change the other, and vice versa.

Value and the environment: Sentiocentrism, biocentrism, and ecocentrism

Valuing animals

If we wish to broaden the circle of moral concern, animals offer a promising starting point. After all, *we* are animals, and we share much in common with many nonhuman animals. To some, we are closely related genetically: scientists estimate that we share more than 98 percent of our genes with chimpanzees (Culotta 2005). Other animals, such as dogs, are closely attuned to human social cues such as pointing, and this is likely due to evolutionary influences associated with domestication (Riedel et al. 2008). We also recognize in animals behaviors and emotional responses we see in ourselves: anthropomorphism—the attribution of human-like qualities to animals—is not always mere projection of human traits onto animals; instead, it can reflect the recognition of traits we share in common. Neuroscientist Robert Sapolsky (1990), for example, found that baboons experience many of the same physiological changes as humans in association with social stress, and philosopher Peter Singer (2002) compellingly argues that we are justified in using a combination of pain behavior, physiological similarities, and evolutionary kinship to conclude that many animals experience pain just as we do.

Our understanding of animals has come a long way since the time of Descartes and his conception of animals as unfeeling machines, or automata. Theoretical developments and empirical evidence from evolutionary biology, cognitive ethology, neuroscience, physiology, and various other fields have rendered a simple Cartesian model of animal life nonviable. Yet despite these significant advances, our ethical practices regarding animals lack guidance by a coherent set of ethical values. We can see this when we reflect on our treatment of animals in various domains. It is widely believed that it would be cruel and morally wrong to leave one's dog in a crate for its entire life, never letting it out for a walk or a romp in the park. But when it comes to pigs, an equally intelligent mammalian species, many of us not only tolerate, but actually support such treatment, buying factory farmed ham for our meals. In the United States, direct disregard for animal life is even more blatant among various groups, such as "varmint hunters" who sometimes kill more than one hundred prairie dogs in a day using long-range rifles, for sport. Yet, cockfighting is illegal. What are we to make of these divergent ways of treating animals? In some realms, it seems that we treat animals as morally considerable and in others, not at all.

Two prominent scholars—Peter Singer and Tom Regan—have played a critical role in calling attention to the inadequacy and inconsistency in our treatment of animals. Both have strongly influenced scholarly debates surrounding animal ethics as well as animal activism and the animal rights movement. Both argue against factory farms and support radical changes in our approach to animal experimentation. However, they represent two different moral outlooks: Singer works in the utilitarian tradition, whereas Regan develops an animal rights theory by extending Kantian notions of respect and inherent value. Despite these differences, Singer and Regan each seek a comprehensive theory that can undergird our ethical understanding and actions in relation to animals. More recently, philosopher Clare Palmer has called into question this approach. Palmer argues that we need to take more seriously the importance of context in defining our ethical obligations to animals. What we may have, then, is not a single grand theory, but various ways of relating ethically to animals, tailored to particular settings. We will examine the views of Singer, Regan, and Palmer in turn.

Peter Singer and *Animal Liberation*

Peter Singer's classic book, *Animal Liberation*, was originally published in 1975 and has been reprinted multiple times since. The book has been a critical resource for the worldwide animal rights movement, and the widely anthologized first chapter, "All animals are equal," has spurred widespread discussion and debate. In the book, Singer throws down the gauntlet regarding our treatment of animals, and by doing so, helps shift the burden of argument onto those who believe that animals are ours to use and dispose of as we wish. Singer's central contributions are twofold. The first is in his defense of equal consideration of the interests of all sentient beings; the second is his detailed reporting on the treatment of animals in agriculture and scientific experiments.

Singer's ethical defense of animals is based on a straightforward moral principle—*the principle of equal consideration of interests*—which, he argues, is both plausible and applicable to animals (Singer 2002, ch. 1). Those who take seriously the interests of human beings, but discount those of animals, are *speciesists*. Speciesism, says Singer, "is a prejudice or attitude of bias in favor of the interests of members of one's own species and against those of members of other species" (2002, p. 6). Singer does not argue that animals and humans need to be treated identically—he is happy to grant that animals should not be given the right to vote—however, he does believe that animals' interests deserve the same consideration and the same weight as similar interests in humans. An animal's interest in avoiding pain, for example, should not be discounted relative to a like interest of a human being.

To the objection that humans are smarter, more powerful, or more sophisticated than animals with respect to our language and culture, Singer says that we do not use such differences to establish different levels of moral consideration among human beings, and neither should we do so with respect to nonhuman animals. We would find it unconscionable to hold one person's suffering as less important than another's on the ground that the first person is less intelligent than the second. Pain is pain, in whomever it occurs, in the weak or strong, foolish or wise, human or animal. Thus, for Singer, if a living being qualifies as having interests—and to do so, it must have "the capacity for suffering and enjoyment" (Singer 2002, p. 7)—its interests are to be counted just the same as those of any other sentient being.

Singer's view is an aggregative, utilitarian one, so it allows interests to be traded off against one another. Because interests sum together, the best course of action, according to this view, may frustrate one (or more) being's interests in order to maximize utility overall. What is not acceptable, however, is to sacrifice critical interests of one being to satisfy the trivial interests of another, for this trades away a large quantity of utility for minimal gains elsewhere. Yet, this is exactly what Singer thinks we do when we subject animals to great suffering in order to satisfy our taste for meat. In principle, it is pain (or frustration of preferences) rather than death that concerns utilitarianism. So Singer acknowledges that under the right conditions, one could eat humanely raised and painlessly killed animals without violating the principle of equal consideration of interests. However, in the real world, the vast majority of the animals that humans consume are neither humanely raised nor painlessly killed. Given this reality, argues Singer, vegetarianism is the moral choice.

Singer offers more than ethical arguments. His second critical contribution has been to carefully document the treatment of animals in agriculture and research, making visible an array of horrifying practices that we support with our tax dollars, our eating habits, and our purchases of cosmetics and other consumer goods. The stories behind the meat on our plate or the moisturizing lotions we apply to our skin are not ones that many of us want to hear, but if we are to take responsibility for our choices, we must grapple with their implications. In *Animal Liberation* (2002), Singer describes the treatment of animals in modern industrial agriculture. Even the most committed carnivore will likely cringe at the description of multiple chickens confined to cages the size of a milk crate, pigs living in stalls where they lack room to turn around, the de-beaking of poultry, and the suffocating ammonia-laden air in contemporary factory farms. *Animal Liberation* is rife with similarly shocking discussions of animal experimentation. Just the sheer numbers are overwhelming: in 1988, the U.S. Department of Agriculture reported that over 1.6 million cats, dogs, primates, guinea pigs, rabbits, and other animals were used in animal experiments (Singer 2002, p. 37). This statistic *excludes* rats and mice that comprise the vast majority of experimental animals (Singer 2002, p. 37). Singer goes on to describe experiments in which animals are force fed cosmetics to determine the level at which 50 percent of the animals die ("lethal dose 50 percent" or "LD50" tests) or suffer prolonged and painful deaths due to repeated electric shocks, intentional overheating, or withdrawal from drugs to which they are forcibly addicted (Singer 2002, ch. 2). These reports suggest that our practices are not in line with our ethical values, that our ethical values themselves need serious readjustment, or both.

Tom Regan and animal rights

Tom Regan agrees with Singer that our moral relationships to animals are deeply problematic. However, he rejects Singer's utilitarian approach. In particular, Regan objects to the tradeoffs that utilitarianism permits. Utilitarianism concerns itself not with individuals per se, argues Regan, but with individuals as containers of value—mere "receptacles" for pleasure and pain:

> For the utilitarian, you and I are like the cup; we have no value as individuals . . . What has value is what goes into us, what we serve as receptacles for; our feelings of satisfaction have positive value, our feelings of frustration have negative value. (Regan 2012, p. 85)

Regan's point is that utilitarianism fails to value individuals *as* individuals. From a utilitarian perspective, a situation in which some cups are empty of utility and others are overflowing is preferable to one in which all cups are equally full, if in the first instance the total sum of utility exceeds that in the second. Since overall utility is all that counts, utilitarianism guarantees no rights to individuals. In Regan's colorful example, if killing your cranky old Aunt Bea would enable you to access her inheritance and distribute it to sick or impoverished children, thereby increasing utility overall, utilitarianism says you should do it. But this cannot be right, argues Regan. We need a theory that begins by valuing individuals.

Regan argues that the fundamental error in our treatment of animals results from our failure to value them as individuals. Our moral relationship to animals cannot be fixed up merely by treating them a bit more kindly or offering them larger cages. What needs to change is our fundamental value orientation toward other conscious beings. As Regan explains, we share in common with other conscious beings the fact that "we are each of us the experiencing subject of a life, each of us a conscious creature having an individual welfare that has importance to us whatever our usefulness to others." (Regan 2012, p. 87). The last part of this quotation has special significance, for it is here that we see Regan's Kantianism emerging: individual animals matter; they are not merely to be seen as means to our ends. Regan's view is that all experiencing subjects of a life have inherent value, and that "all who have inherent value have it equally, whether they be human animals or not" (Regan 2012, p. 87). On this basis, argues Regan, we cannot sustain current systems of animal agriculture or scientific experimentation with animals, because these practices do not respect animals as individuals with inherent value equal to our own. Only total abolition of these practices can adequately address "the fundamental wrong" which is seeing animals as resources, to be used and disposed of as we wish (p. 81). Regan describes his view as "the rights view," because he asserts that not only humans, but all "experiencing subjects of a life" have equal inherent worth and hence "equal right to be treated with respect" (p. 87).

Clare Palmer's ethical contextualism

Regan's theory differs from that of Singer's in two key ways. First, it asserts that each individual subject of a life has a right to respect in light of its inherent worth, and respect is not something that can be traded away to advance the general welfare. Second, whereas Singer focuses on sentience and the capacity to suffer, Regan seems to rely on a more robust notion of consciousness: subjects of a life "want and prefer things; believe and feel things; recall and expect things" (p. 87). Thus, Singer (2002) has suggested that from a taxonomic perspective, moral consideration may end somewhere in the vicinity of simple mollusks such as oysters (though we cannot be sure of this), whereas Regan (2004, p. xvi) suggests that we can be certain that mammals over a year of age are subjects of a life, and other kinds of animals may be as well. Perhaps both views are limited, however. Clare Palmer suggests that what she calls "capacity-oriented accounts of animal ethics" are limited by their exclusive focus on the capacities of animals for suffering, consciousness, and the like (Palmer 2010, ch. 2). What such accounts fail to recognize is the significance of our *relationships* to animals in determining our obligations toward them. By taking these relationships into consideration, we can better understand why people often feel that they have obligations to protect and care for domestic animals, while we lack such obligations to animals in the wild (Palmer 2010, ch. 4).

Palmer labels the idea that we have limited obligations to care for wild animals the "laissez-faire intuition." But why might we have weaker obligations to wild animals than to their domestic counterparts? In reply, Palmer argues that our engagement with animals can generate special obligations to assist them. In the case of domestic animals, human beings significantly shape their lives and welfare: we have bred domestic animals to be dependent on us, for example, and in most cases, it is human actions that make possible their very existence (Palmer 2010, p. 91). With wild animals, similar relationships of dependency typically do not exist—though, as Palmer (2010, ch. 5) notes, our entanglements with wild animals can be complex. Human-caused global climate change, for example, now threatens the existence of polar bears, yet it is not easily possible for us to immediately withdraw this harm. Thus, Palmer suggests that we aid polar bears through other means—such as protecting their habitat from other forms of damage. Although Palmer's relational approach does not dictate a clear answer in every case, it can provoke a more nuanced and differentiated understanding of what we owe to animals. What utilitarian and rights approaches overlook, suggests Palmer, are the ways in which particular relationships can generate special obligations. This is a point that many readily acknowledge in the case of human relationships: becoming a parent, for example, generates significant moral obligations, and one's obligations to one's own children are much stronger than one's obligations to children in general. Palmer's ethical contextualism explains how our obligations to animals might be similarly relationally dependent.

Valuing life

Although Palmer questions the ability of capacity-oriented accounts to capture all the relevant dimensions of animal ethics, the general strategy of identifying morally significant capacities, traits, or qualities has been central to both animal and environmental ethics. From the perspective of environmental ethics, however, a focus on conscious or sentient beings may not go far enough. Think about what kinds of values environmental concerns often involve. Concerns about clean water and air focus primarily on human needs and interests, and people put bells on their cats to spare the lives of individual songbirds, but those worried about biodiversity, endangered species, or the protection of majestic old-growth forests have concerns that go beyond those at the heart of either anthropocentric or sentiocentric theories. In some cases, in fact, we find conflicts between those primarily concerned with animal rights and those focused on environmental issues more broadly. In Olympic National Park, for example, park officials in the 1980s began working to remove mountain goats that were damaging alpine plant communities. Animal rights groups—such as the Fund for Animals—protested (Scheffer 1993). Similarly, efforts to control burgeoning populations of deer and geese through hunting are often supported by environmentalists but opposed by advocates of animal rights. Although there is often consensus on other issues—such as the moral problems with factory farms—many philosophers believe that sentiocentric views do not sufficiently expand the circle of moral concern.

In a classic essay, Kenneth Goodpaster (1978, p. 310) puts the point like this:

> Neither rationality nor the capacity to experience pleasure and pain seem to me necessary (even though they may be sufficient) conditions on moral considerability. And only our hedonistic and concentric forms of ethical reflection keep us from acknowledging this fact. Nothing short of the condition of being alive seems to me to be a plausible and nonarbitrary criterion.

Goodpaster suggests that theories centered on rationality or sentience exhibit bias in favor of our particular form of life, failing to recognize the fundamental source of moral value: being alive. Central to Goodpaster's argument is the idea that the characteristics we take to be critical conditions for moral consideration have value only derivatively, as means to the end of promoting life. Goodpaster (p. 316) explains:

> Biologically, it appears that sentience is an adaptive characteristic of living organisms that provides them with a better capacity to anticipate, and so avoid, threats to life. This at least suggests, though of course it does not prove, that the capacities to suffer and to enjoy are ancillary to something more important rather than tickets to considerability in their own right.

Thus, although plants may not *cognize* their interests, they are organized, homeostatic systems oriented toward self-maintenance, and in this sense, they have an interest in staying alive and healthy.

To embrace the possibility of a life-centered, or biocentric, ethic would require a substantial shift in dominant worldviews. Paul Taylor (1981) thinks that such a shift is possible, and he argues for an attitude of respect for nature, supported by a belief system that he calls "the biocentric outlook on nature." Fundamental to this belief system is a de-centering of human beings and their importance in the overall scheme of life on earth. Taylor encourages us to think of ourselves as just one form of life—and a relatively young and late arriving one—in a stunningly diverse array of species with which we share the planet. What's more, we are dependent on other forms of life: the natural world is a "complex but unified web of interconnected organisms, objects, and events" (p. 209). Lastly, Taylor argues that each living thing is "a teleological center of life, pursuing its own good in its own way" (p. 207) and we should recognize and value those diverse goods. Taylor explains, "When considered from an ethical point of view, a teleological center of life is an entity whose 'world' can be viewed from the perspective of *its* life" (p. 211). Although accepting these ideas does not logically *entail* that we reject human superiority, Taylor suggests that the first three elements of a biocentric outlook show a commitment to the superior value of human beings to be "an irrational bias in our own favor" (p. 216). Once we realize this, we are open to seeing ourselves in moral relationship to all living things. Just as we seek to pursue our good, they seek to pursue theirs. Just as humans have value as living things, so do members of other species. These insights in turn ground an attitude of respect for nature: recognizing the inherent value of all living things prompts us to see that all living things are deserving of respect.

Taylor here tries to achieve a gestalt shift in our conception of ourselves in relation to the living world. Rather than see ourselves—*Homo sapiens*—as sitting atop a Great Chain of Being, we can understand ourselves as members of an interconnected web of life. We are by no means the same as other members of that web, but we are not superior to them either. Because Taylor recommends a wholesale shift in the dominant worldview, he cannot easily argue from a set of incontrovertible shared premises to a conclusion that we must accept on pain of inconsistency. Nevertheless, he adduces evidence—highlighting our short evolutionary history, our dependency on other elements of the living world, and the unique capacities of other organisms that equip them well for their own modes of life—to show the plausibility of the view he espouses.

This methodological point is important. As the positions of environmental philosophers move from the conservative to the radical, it becomes increasingly difficult to support one's conclusions with deductive arguments from accepted premises. This should not be taken to imply that arguments for more radical conclusions need not be thoughtful or well reasoned, but that the task may be somewhat differently defined. Those who advocate substantial shifts in our moral outlook need to envision and describe alternatives, and to show why these alternatives make sense. But their arguments may engage us in different ways: by prompting us to envision our place in the world differently, asking us to imaginatively try on or test out different ways of thinking and acting (see Fesmire 2003).

This may be where philosophy extends into other modes of thought and communication, such as literature and the arts. For although formal arguments for biocentrism may be relatively few in number, we frequently encounter the modes of valuation recommended by biocentrists in the writings of those closely engaged with the natural world. Awe and respect for other forms of life are apparent in the work of writers such as Albert Schweitzer, E. O. Wilson, Rachel Carson, Henry David Thoreau, Willa Cather, Konrad Lorenz, Mary Oliver, Aldo Leopold, and many others. Close attention to and careful observation of the natural world often help trigger an understanding of other living things as pursuing their own goods in their own unique and fascinating ways. For this reason, a shift to the biocentric ethic recommended by Goodpaster and Taylor may require not only philosophical argument, but greater engagement in the natural world.

Valuing ecosystems, species, and biodiversity

Thus far, the extension of moral concern has remained focused on individuals: individual animals or individual living things. Taylor and Goodpaster allude to the possibility that their views might extend to the valuation of larger ecological wholes, but neither Taylor nor Goodpaster develops this possibility in any depth. Intuitively, it may not be surprising that ecological systems, species, or biodiversity should have value in their own right. It has been suggested that beauty, wisdom, democracy, freedom, peace, friendship, and many similarly abstract entities have intrinsic value, and if this is right, then the realm of intrinsic value can encompass much more than individual organisms.

Many arguments for the protection of biodiversity cite its intrinsic value (see, e.g., Ehrenfeld 1988), and Aldo Leopold is known for the idea that the biotic community has value in its own right. Moreover, environmental ethics may *need* an account of the value of ecological wholes to capture important aspects of environmental concern. When people express concern about the decline of coral reefs, for example, their concern is neither exclusively about the physical structure of the reef nor about the individual organisms that occupy it. Rather, people worry about the future of coral reefs as *living systems*, valuable not only for the elements that comprise them, but for the complex interplay of those elements over time.

Despite all this, the valuation of ecosystems, species, and biodiversity has been a difficult sticking point in environmental philosophy. Ecocentric theories—those which hold that ecosystems are morally considerable—have been variously criticized as unintelligible, misanthropic, and even fascist

(Bookchin 2002; Cahen 2003; Regan 2004). The question is, why? We will examine this question in more detail below, but for now, it is useful to recall the centrality of *individualism* in Western moral philosophy. From Aristotelian virtue ethics to Kant's emphasis on the autonomy of rational beings to Thomas Hobbes' view of ethics as a contract among rational, self-interested persons, individuals have been the central focus of our moral attention. In many traditional philosophical theories, it is both individuals who can be moral agents and who are the subjects of moral concern. Typically, individuals are held to be the subjects of moral concern in virtue of some quality or capacity they possess: rationality, sentience, having interests, or being alive. None of these qualities neatly applies to ecosystems, species, or biodiversity.[1] For this reason, the arguments given in favor of an ecocentric ethic typically: (a) attempt to extend one or more of these qualities, such as having interests, to ecological systems, or (b) appeal to novel features of ecosystems as the source of their value.

As an example of the first approach, we might take scientist James Lovelock's Gaia hypothesis. In his 1979 book, *Gaia: A new look at life on Earth*, and in many subsequent writings, Lovelock argues that the earth as a whole is a living system, of which we are a part. Although his ideas are controversial, and he does not make an explicit philosophical argument for ecocentrism, the earth-as-organism metaphor could be used in developing an analogy between ecological wholes and the organismic individuals who are the paradigmatic subjects of moral consideration. If the earth is an integrated, self-maintaining living system, then it presumably has the kind of goal-directedness that Paul Taylor considers key to having inherent value. If Lovelock is right, then the earth itself is (in Taylor's phrase) "a teleological center of life."

Harley Cahen (2003), however, resists the idea that ecological systems share key features with individual organisms. In particular, he argues that ecosystems are not goal-directed and hence, they do not have interests. In schematic form, here is his argument:

1. To be morally considerable, *x* must have interests.
2. To have interests, *x* must be goal-directed.
3. Ecosystems are not goal-directed.
4. Therefore, ecosystems do not have interests.
5. Therefore, ecosystems are not morally considerable.

The central point of Cahen's argument is to show that interests require goal-directedness, and that ecosystems lack such teleological organization. On the connection between interests and goal-directedness, Cahen argues that it is goal-directedness that makes it possible for a living thing to have a standpoint, enabling us to assign interests to it. Goal-directedness, while not sufficient for moral considerability, is at least necessary for it, and ecosystems are not goal-directed. This is because, Cahen argues, the features of ecosystems that seem to indicate goal-directedness—structure, patterns, or stability, for example—are mere *byproducts* of the individual elements that comprise them. There is no ecosystem-level homeostatic mechanism that holds the whole system together. Ecosystems are not organism-like; instead they are an amalgamation of organisms of various species, along with their abiotic environment. The interactions among these organisms and their environment may produce certain patterns, but an ecosystem itself is not a goal-directed, homeostatic system, and any appearance of such is mere illusion.

There are a variety of ways to respond to this argument. First, one might wonder whether Cahen is entirely right that ecosystems lack any kind of teleological organization or systemic unity. Second, one might question whether organism-like integration is necessary for an ecological system to be valuable, or to have a "good of its own" that we can and should take into account. Ecocentric theorists have employed both strategies in developing their views. Holmes Rolston III and J. Baird Callicott, for example, suggest that it would be a mistake to expect ecosystems to conform to the paradigms of moral considerability we employ for human beings, sentient animals, or even individual living organisms more generally.

Callicott's work takes as its touchstone the writings of Aldo Leopold, and in particular, his land ethic. Leopold, a naturalist, forester, wildlife biologist, and critical early figure in ecological restoration and the American wilderness movement, provided an important seed for the development of ecocentrism in environmental philosophy. Leopold's most philosophical writings are contained in *A Sand County Almanac* (1949), in which he reflects on his experiences in nature, from his time working for the Forest Service in the southwestern United States to his efforts to restore an old, degraded farm in Wisconsin. The final section of the book, entitled "The upshot," contains a collection of more formal essays focused on wilderness, wildland recreation, wildlife, and the land ethic.

Leopold (1949, pp. 224–225) is perhaps best known for a dictum he offers near the end of that final essay: "A thing is right when it tends to preserve the integrity, stability, and beauty of the biotic community. It is wrong when it tends otherwise." This dictum suggests that the biotic community itself has moral status, and that its integrity, stability, and beauty should guide our moral decision-making. Indeed, the essay as a whole works to establish the view that our moral community should no longer end at the boundaries of humanity, but extend beyond it "to include soils, waters, plants, and animals, or collectively: the land" (p. 204). It is clear here and throughout Leopold's essay that he intends us to think not only about individual organisms as worthy of moral attention but also to value living systems, which include organisms of various species along with their abiotic components. Although Leopold himself does not use the term, *ecosystems* seem to be at the heart of his concern. "Land," for Leopold, denotes this idea: for example, he describes the land as "a fountain of energy flowing through a circuit of soils, plants, and animals" (p. 216).

Leopold did not use terms like "moral considerability" or "intrinsic value"; he left it to his successors to reconstruct his arguments in more formal philosophical terms. J. Baird Callicott took up this task, acting as both an interpreter and defender of Leopold's views. Callicott (1987) suggests that Leopold's ethical views grow out of Humean-Darwinian theoretical foundations, and that David Hume, Adam Smith, and Charles Darwin think of moral value not as "objectively present in people and/or animals" but instead "projected by valuing subjects" (Callicott 1987, p. 198). What matters, on this view, is not intrinsic value as a property of ecosystems in a deep metaphysical sense, but whether *we* ought to value ecosystems instrinsically, for their own sake (Callicott 1999a). By making this shift, Callicott focuses our attention on modes of human valuing, rather than on what possesses value independent of valuing subjects. Using this approach, he goes on to argue that *it makes sense for us to value ecosystems* in their own right.

Callicott's argument for this point draws on the science of ecology, which, he argues, shows us the importance of ecological wholes. In ecosystems, individuals are what they are by virtue of their place in the system: ecology shows us that systems and relationships have ontological priority over individuals. Callicott thinks this in turn supports the view that ecological systems and ecological relationships should be *ethically* valued. As Joseph DesJardins (2006, p. 198) puts it, "Callicott reasons from ecology to metaphysics to ethics." Whether this move succeeds is a matter of contention, but Callicott can, at least, shrug off Cahen's objection that ecosystems are not morally considerable because they lack interests. Callicott's argument for valuing ecosystems does not depend on the claim that ecosystems have interests.

There is another important objection, however, that Callicott faces. Callicott (1987, p. 196) argues that the land ethic "not only has a holistic aspect; it is holistic with a vengeance." Claims like this have prompted the objection that the land ethic is a form of "environmental fascism": by subordinating individuals to the good of the whole—as Leopold's famous dictum seems to suggest—the land ethic has no place for the protection of individual rights (Regan 2004, pp. 361–362). If, for example, the integrity, stability, and beauty of the biotic community would be vastly improved through a program of mandatory human sterilization, would the land ethic endorse it?

Callicott counters the fascism objection by arguing that the land ethic is meant to overlay other ethical commitments, serving as an addition to them, not a replacement. He acknowledges that there may be conflicts between duties to individuals and duties to larger wholes, but suggests that these conflicts can be worked out by appeal to second-order principles that allow us to prioritize: in general, duties to those close to us take precedence over those far away, and stronger interests take precedence over weaker ones (Callicott 1999b). Fleshing out and defending these principles is a task in its own right, but Leopold's text supports the point that the land ethic is meant as an addition to our existing ethical commitments. Leopold argues throughout his essay that ethics develops through a gradual outward expansion, and that expansion to encompass the land is the next logical step.

Ecocentrism, then, need not imply that ecosystems are the only things with moral value, or that the protection of ecosystems always trumps other moral concerns. Ecocentric theories typically have a nested structure, where wholes *and* parts count morally. Holmes Rolston's moral outlook shares this structure. Unlike Callicott, Rolston (1988) holds on to objective value, arguing that value existed in the world prior to the presence of human valuers. Rather than work to find similarities among individual organisms, species, and larger ecological wholes, Rolston urges us to recognize that value takes diverse forms. There is no single characteristic, whether rationality, sentience, or being alive, that all valuable things share in common.

In the case of plants and other nonsentient beings, we find value in their teleological organization: they are entities that "defend value" (Rolston 2012, p. 100). By this, Rolston means that living organisms are organized systems that seek to protect this order against threats. They are self-maintaining, goal-directed systems. In species, we find "dynamic life forms preserved in historical lines that persist genetically over millions of years" (Rolston 2012, p. 129). Thus, to kill a species is to "[shut] down the generative process, a kind of superkilling" (Rolston 2012, p. 135). Ecosystems, too, possess distinct forms of value. They have "no brain, no genome, no skin, no self-identification, no *telos,* no unified program" and they lack interests (Rolston 2003, pp. 148–149). However, ecosystems *produce* value: they generate complex order and interdependencies, they foster adaptive fitness, and they are the source of a stunning array of valuable organisms. Rolston thus argues that ecosystems have what he calls "projective value," which is a kind of generative value associated with the production of life in all its diverse forms: "Nature is a fountain of life, and the whole fountain—not just the life that issues from it—is of value" (Rolston 1988, p. 197). We find this idea echoed more simply in another essay, where Rolston (2003, p. 150) tells us that "[i]t would be foolish to value the golden eggs [e.g., the myriad species that ecosystems produce] and disvalue the goose that lays them."

However, not everyone finds Rolston's account of generative value convincing. In particular, it has been objected that Rolston commits the *genetic fallacy* by arguing that ecosystems (and nature as a whole) have value because they produce things of value. As Bernard Rollin (1995, p. 56) notes, "The cause of something valuable . . . is not necessarily itself valuable." Yet Rolston would likely push back against this objection, arguing that although Rollin is correct in principle that something valuable might be caused by something without value, in this case the sheer generative power of nature *is* valuable. It is not as if this generative power is mostly bad, but incidentally produces some good: nature is fundamentally valuable as the source of the further value contained in organisms and living systems. Rolston faces other challenges, however. For example, he must establish that the projective value of nature is a kind of *moral* value, and that this value generates duties (see McShane 2007a, p. 5).

The lesson we can take from this discussion is that it is not easy or straightforward to establish the intrinsic moral value of ecosystems, and thereby ground an ecocentric ethic. Intuitively, many people find it morally problematic to destroy an entire ecosystem—but is this because the ecosystem *itself* has moral value? Or is it that the ecosystem comprises valuable *parts*, whose continued existence and flourishing would be destroyed along with the ecosystem? Ecocentric views appeal to the idea that ecological systems *themselves* have value, above and beyond the value of their individual parts. When Leopold talks about the integrity, stability, and beauty of the biotic community, he is referring to system-level properties. Just as the beauty of a novel is not determined by the beauty of the individual words it contains, so too the beauty of the biotic community cannot be determined by adding together the beauty of each of its elements. It is instead a property of the whole, which is determined not only by the individual parts but by the relationships among them.

The relationship between parts and wholes raises many interesting and challenging philosophical questions, but to address them here would bring us too far afield. Suffice it to say that the challenges in establishing an ecocentric ethic involve not only questions of ethics, but questions of metaphysics, epistemology, and ontology: What is an ecosystem? How is an ecosystem related to its component parts? How can ecosystems be individuated—that is, how can we determine the boundaries of an ecosystem, or where one ecosystem ends and another begins? Can we really value ecosystems if we cannot define or individuate them? Even though these questions are difficult to answer, they don't decisively undermine the prospects for a viable ecocentrism. Nevertheless, one might reasonably wonder at this point whether the traditional approach of focusing on intrinsic value and moral considerability is the most constructive one. Perhaps, we should shift the conversation from the metaphysics of value to the quality of our relationships with animals, plants, and the natural world as a whole.

For further thought

1. How might Singer and Regan each approach the question of whether or not it is permissible to raise and kill animals for the sole purpose of human consumption?
2. In what cases might the views of animal rights theorists, biocentrists, and ecocentrists conflict? Give an example.
3. What are the advantages and limitations of a contextualist approach to ethics, such as that of Clare Palmer?
4. Can ecosystems have interests? Can they be morally considerable? Critically discuss Harley Cahen's argument.

Beyond intrinsic value? Relational approaches to ethics

Efforts to extend value to animals, all living things, and ecological wholes have drawn significant attention to the narrowness and human-centeredness of traditional approaches to morality, and they have been crucial in broadening our moral awareness. However, after decades of effort, many scholars remain unsatisfied with the results of the extensionist approach. The concerns are diverse. Some worry that sentiocentrism, biocentrism, and ecocentrism fail to offer viable alternatives to anthropocentrism because they do not adequately account for competing values and conflicting moral claims. Alan Carter (2011), for example, argues that audacious claims about the equal inherent worth of all living things (or all experiencing subjects) do not hold up well in the practical realm, where we regularly face tradeoffs and cannot avoid choices about whose lives and interests to prioritize. Such tradeoffs have led theorists such as Tom Regan to supplement their accounts with additional principles designed to facilitate prioritization. But such principles often seem ad hoc—they do not follow from the original theory and in some cases may conflict with it. Other approaches, which sidestep moral conflicts by prioritizing a single value, such as the integrity of ecosystems, simply seem implausible (Carter 2011). Surely individual human beings, at least, have value in their own right, independent of their contribution (or lack thereof) to the integrity of ecosystems. Carter argues on the basis of these observations that we need a value pluralist environmental ethic: one that systematically takes into account multiple values—of humans, animals, and ecosystems.

Others have argued for value pluralism as well, but with less optimism about the prospects for a systematic, unified theory. Environmental pragmatists such as Andrew Light, Bryan Norton, Anthony Weston, and others have argued that we value living things and the natural world in diverse ways, and we should neither reduce ethics to a single value, nor do we need a single, foundational account of what possesses moral value. Instead, we should focus on public dialogue that can generate consensus on matters of environmental policy. Norton (1991), for example, suggests that even with diverse and divergent underlying values, we can often come to agree on practical matters involving the environment, thus debates over intrinsic value are unnecessary.

Katie McShane (2007b), however, argues that we need the concept of intrinsic value, because it captures a distinct and important *way* of valuing the natural world. McShane acknowledges that intrinsic value has been understood in various ways by environmental ethicists. Although not all concepts of intrinsic value may be essential, at least one should be retained. In particular, we need intrinsic value claims to call attention to "the distinctive way that it makes sense to care about [certain elements of the natural world]" (2007b, p. 47). The idea here is that intrinsic value claims are claims about *the kinds of attitudes that are appropriate* in response to particular aspects of nature. An example may help clarify. Think of the way you value various things: your parents, your friends, your bicycle, trees around town, cell phones, pizza, and so on. It is likely that you have somewhat different attitudes toward each of these things, and the differences are warranted. If you valued pizza and your parents in exactly the same way, you would arguably be confused. Even things we value similarly—like parents and friends—warrant slightly different attitudes. We owe our parents a different form of respect than we owe our friends, for example.

McShane's point is that certain, important moral attitudes are intrinsically valuing attitudes: when we value something intrinsically, we value it "for its own sake, or in its own right" rather than "for the sake of some other valuable thing" (p. 50). Love is an intrinsically valuing attitude, because "to love something is [in part] to value it as a good in itself" (p. 52). Similarly, reverence is an intrinsically valuing attitude. We cannot revere something without valuing it in its own right.[2] Thus, the idea of intrinsic value allows us to capture the point that certain things merit intrinsically valuing attitudes: the Grand Canyon might warrant an attitude of awe and respect, and a rocky intertidal zone teeming with life might warrant an attitude of wonder.

Notice that McShane's account of intrinsic value does not emphasize metaphysics or ontology. She neither argues that animals, plants, or the natural world have value independent of any human valuer, nor does she argue that value is entirely subjective. Instead, she claims, it makes sense to value certain things in certain ways. This shift toward valuing attitudes resonates with the relational approaches to environmental ethics explored below. Relational approaches emphasize the nature and quality of our relationships as loci of value—and certainly our valuing attitudes are important aspects of our relationships, whether with people, animals, plants, artifacts, or ecosystems. Although relational approaches have not been well developed as an alternative to extensionism and pragmatism, there are many resources on which to draw in elaborating a relational environmental ethic.

Leopold's relational ethics

The work of Aldo Leopold is a good place to start in considering the potential for an ethic that makes relationships central. Although Leopold's famous dictum regarding the integrity, stability, and beauty of the biotic community has garnered significant attention for its ethical significance, this single line is not necessarily the most important ethical insight in his work. More significant, perhaps, is the way in which Leopold exemplifies sensitive relationships to the natural world and an empathic connection to the land. If we focus not only on "The land ethic," but return to *A Sand County Almanac* more broadly, we find numerous passages in which Leopold is deeply involved in close, careful, and emotionally engaged observation of the world around him. Throughout the book, Leopold watches and listens, trying to understand the creatures he encounters: What are they doing, and why? How are they feeling? What might it be like to be a tree, a goose, or a mountain? Though he does not always succeed, Leopold works to connect with animals and plants on their own terms, and to see himself as a "plain member and citizen" of the biotic community, as he later advocates in "The land ethic." Take the following passage from the opening section of *A Sand County Almanac*, "January thaw":

> A meadow mouse, startled by my approach, darts damply across the skunk track. Why is he abroad in daylight? Probably because he feels grieved about the thaw. Today his maze of secret tunnels, laboriously chewed through the matted grass under the snow, are tunnels no more, but only paths exposed to public view and ridicule . . .
>
> The mouse is a sober citizen who knows that grass grows in order that mice may store it as underground haystacks, and that snow falls in order that mice may build subways from stack to stack . . . To the mouse, snow means freedom from want and fear. (Leopold 1949, p. 4)

Leopold engages with the natural world as a sensitive person engages with other human beings, by attempting to understand other individuals in all their uniqueness and particularity, with careful attention to context. He draws analogies between the familiar human realm and the natural world he is observing, not as a way of imposing his own experience onto nature, but as a way of reaching out and trying to connect. This again, is something we do in our relationships with other human beings. For example, although we would be mistaken to assume that a friend's experience of disappointment is just like our own, we may use our own experience as a way to begin to understand and explore theirs. In conversation and through attentive listening, we can better comprehend *their* disappointment: the analogy between ourselves and others helps get us started, and careful attention helps us to see more fully their point of view.

Deep ecology and the relational self

Perspective taking plays a similarly important role in deep ecology, an environmental outlook introduced by Norwegian philosopher Arne Naess. Deep ecology is not itself a normative ethical theory, but instead a worldview that emphasizes our connectedness to other living things and the possibility of living in a way that fosters the flourishing of all life. Naess sees *self-realization* as a critical element of a deep ecological worldview, and for him, self-realization requires an expanded conception of the self that extends far beyond the narrow ego. This process of self-realization involves identification with other living things, which Naess describes as a "process through which the interest or interests of another being are reacted to as our own interest or interests" (Naess 2012, p. 136). Naess offers an example of identification based on his own experience. Watching a chemical reaction under a microscope, Naess sees a flea fall into the acid:

> To save [the flea] was impossible. It took many minutes for the flea to die. Its movements were dreadfully expressive. Naturally, what I felt was a painful sense of compassion and empathy . . . (Naess 2005, p. 518)

Naess suggests that his ability to empathize with the flea is grounded in his identification with the flea—his ability to project himself into the flea's position. The details of identification are complex and controversial: Can one really imagine oneself to be a flea? Or is identification more a matter of putting oneself in the position of the flea, thinking that since an acid bath would be bad and painful for oneself, it must be similarly bad and painful for the flea? However, the basic idea is that taking the perspective of others allows us to develop a richer understanding of and compassion for all living beings.

Identification is also tied closely to our identities, to our sense of self. Naess argues that our identities are constituted through our relationships with humans and nature:

> We may be said to be in, and of, Nature from the very beginning of our selves. Society and human relationships are important, but our self is

> much richer in its constitutive relationships. These relationships are not just those we have with other people and the human community. (Naess 2005, p. 516)

For example, when people relocate from one place to another—particularly when they have lived in a place for some time—they experience a "loss of personal identity," according to Naess (2005, p. 521). The places we live enter into and shape who we are: people and places are woven together.

Deep ecology explicitly focuses on self-in-relation. We are not isolated and autonomous thinkers, knowers, or actors, and though we are part of a connected and interdependent web of life whether we realize it or not, deep ecology enjoins us to embrace our interconnectedness through the process of identification, and by expanding our self-conception to encompass all life. This, deep ecologists suggest, can serve as the foundation for significant social and political change. If we connect and identify with all living things, we will be motivated to live more simply and lightly, to embrace life quality over life quantity, and to recognize the need to live within ecological limits in order to make possible the ongoing flourishing of many forms of life (cf. Naess and Sessions 1995). Deep ecologists thus see a strong connection between the personal and the political, where the development of a wider, ecological self can catalyze a commitment to "deep questioning," a searching, critical exploration of the assumptions that drive our current political and economic systems, and a move to reorient away from materialism and infinite growth toward a more holistic, life-centered view.

Deep ecologists frequently cite Asian traditions of thought in an effort to jar loose the individualistic conceptions of the self that dominate in the West. Although deep ecologists have been criticized for appropriating Eastern traditions such as Buddhism and Daoism for their own ends (see Guha 1989), there are connections between a number of Eastern traditions and a more relational conception of the self. These connections are worth exploring, though we should be careful not to assume that the relational self of Confucianism or the expansive sense of perspective characteristic of Daoism is identical with the citizen of the biotic community championed by Leopold or the realized self of deep ecology. The Confucian relational self, for example, is not an *ecological* self. Nevertheless, there are conceptual resources in Confucianism and Daoism that may be helpful in developing a relational approach to environmental ethics.

Classical Confucian and Daoist perspectives on the relational self

The Confucian tradition originated in China, but has played an influential role throughout much of East Asia. The Classical Confucian tradition encompasses

Confucius himself (see Lau 1979), as well as the important but less well-known philosophers Mengzi (Mencius) and Xunzi. The self, in this tradition, is constituted relationally. We are who we are in virtue of our relationships with others. As Henry Rosemont, Jr. (1991, p. 90) put it, "[T]here can be no me in isolation, to be considered abstractly: I am the totality of roles I live in relation to specific others . . ." Yet what does this mean? Routley's last man case, though fictional, seems conceptually possible. Surely I would remain myself, even if all other human beings were extinguished!

In a biological sense, it is of course possible for a single human being to exist on earth (at least for a time). But from a Confucian perspective, that human being could not be a *person* in the most robust sense of the word. This is because the Confucian conception of personhood is a moral one, and personhood can only be achieved in relation. In the Confucian tradition, *ren*, or humanity, is the central virtue, and to be *ren* is to engage well with others, with care and sympathetic understanding. Interactions with others are structured by tradition and ritual, which provide the means to express moral attitudes such as respect, gratitude, or reverence. However, the humane person possesses a kind of moral virtuosity: she can navigate gracefully through morally challenging situations, responding with compassion and sensitivity. This is not to say that a person who is *ren* never provokes anger or resentment in others, or that she can prevent all suffering and harm, for this is clearly impossible. But the humane person, or *junzi*, is one whose relationships with others are uniquely sensitive to the context and the particularities of those with whom he interacts.

The *junzi*'s relationships both exemplify and generate harmony among persons. In this way, the virtue of *ren* is tied to the overarching notion of the *dao*, or way. The dao is a normative ideal, representing something like harmonious existence. To be *ren* is to act with care and attentiveness, in ways that promote harmony, but the *ren* person acts in *accordance* with the *dao*, not just in ways that *promote* the *dao*. The distinction is a fine one, but understanding it can help us see how the *dao* functions as an ideal. The *dao* is more like *eudaimonia* in Aristotelian ethics than it is like happiness in utilitarianism. In utilitarianism, one seeks to promote certain states of affairs. In Aristotelian virtue ethics, one seeks to *live* virtuously, to *be eudaimon*. The *dao*, similarly, is something that one seeks to embody and to live, rather than to promote. In a sense, the *dao* is already present and what one does by acting well is to align with the *dao* and to amplify it. To use a musical analogy, virtuous action for Confucians involves something like taking up a part in an ongoing symphony. The challenge is to blend with what is already there, while also drawing it out through harmonious engagement.

For Confucians, the foundations of respect and gratitude develop in the family, as children recognize their dependency on others and the centrality of caring and respectful human relations. When children experience sympathy, care, and love from their parents, they develop their own capacities for care and love. In addition, a child's recognition of dependency on his or her parents lays the foundations for a lifelong debt of gratitude and the virtue of filial piety, which involves respect and deference to parents and elders. Whereas the dominant conception of the mature person in the West is the autonomous individual who has overcome dependence, the recognition of interdependence as an essential and ongoing feature of human life is critical to Confucianism, and to Chinese philosophy more generally (see Lai 2006).

The classical Confucian philosophers – Confucius, Mengzi, and Xunxi – each offer a slightly different ethical outlook, but in all cases sympathetic concern and care are central, as is the idea that morality inheres in the quality of our relationships with others. Although the Confucian relational self is grounded primarily in *human* relations, the idea that our interactions with others both constitute our identity and are at the heart of what it is to be a fully developed person may provide a fruitful direction for environmental ethicists to pursue (Hourdequin and Wong 2005). In extending a relational approach beyond human beings, it is worth exploring early Daoist thought, particularly the work of Zhuangzi.

Although less frequently read than his counterpart Laozi (putative author of the *Dao de Jing*), Zhuangzi's work offers a provocative and sometimes jarring challenge to conventional ways of valuing the world. Whereas Confucianism emphasizes virtues such as compassion, respect, and filial piety, which find expression through tradition and ritual, Zhuangzi seems to think that tradition and ritual can obscure the *dao* by limiting our perspectives. Instead, he stresses the importance of attentiveness to what contemporary philosophers might describe as "the things themselves." This attentiveness, in turn, is made possible when assumptions about the world, our modes of classifying things, and our ways of valuing relax their grip.

Millienia before Kafka's *Metamorphosis*, Zhuangzi asks us to imagine the transformation of a fish into a giant bird that soars high above the earth, to the amazement of smaller creatures like a cicada and a dove. We are invited to consider how different the world looks from the perspective of the big versus that of the small. By playing with perspective, Zhuangzi calls on us to loosen our own attachment to a fixed way of seeing things. He mocks Huizi, who cannot think of any use for things that fail to fit his conventional expectations. When Huizi smashes some giant gourds, deeming them useless because they wouldn't work as ladles, Zhuangzi replies:

> You, sir, are certainly clumsy about using big things . . . Now you had these gigantic gourds. Why not lash them together like big buoys and go floating on the rivers and lakes instead of worrying that they were too big to dip

> into anything? Your mind is full of underbrush, my friend. (Zhuangzi 2001, pp. 212–213)

Ultimately, Zhuangzi seems to be working to open our minds to the *possibilities* present in the world, possibilities open to us if we can stop imposing on things our narrow ways of thinking, which blind us to what is really there.

For Zhuangzi as for Confucius, the virtuous person is in tune with and guided by the *dao*. But the Daoist *dao* differs from the Confucian one in important ways. It is a broader, more cosmopolitan *dao*. It encompasses not just people, but the natural world. It is not about hands-off reverence for nature, but about seeing nature in a broader, richer, and more attentive way, and not just in terms of what it can provide for us economically or materially. The ideal person, for Zhuangzi, not only *comprehends* the world aright, but acts in tune with it. Here we encounter the idea of *wu-wei*, or "nonaction." But wu-wei is not refraining from action, it is interacting well, interacting so harmoniously that action is not imposed upon but rather follows along with its interactants. Ivanhoe and Van Norden (2001, p. 393) describe wu-wei as "effortless action." As they explain, "nonaction is acting in a way that is natural, unforced, and unself-conscious" (p. 393).

Returning to environmental philosophy, we might extend and supplement the Confucian relational perspective with a Daoist one. On this view, relationships remain morally central, but relationships are not confined to the human-to-human. Instead, we cultivate ourselves as moral persons by developing sensitive relations with people, animals, trees, and the world as a whole:

> We are not only members of this or that family, and this or that community, but interdependent citizens of the world, and we must act appropriately in the largest circles, as we must in the smallest. To become good citizens of the world, we need not exchange the particular for the general, abandoning what is learned in relationships to local people and places, but rather learn to take up diverse points of view, in all their richness. A cosmopolitan relational identity is one in which the individual is sensitive to a diverse array of particularities . . . Within the largest circles of concern, the Confucian relational self becomes a Daoist relational self who makes the whole world her home. (Hourdequin and Wong 2005, p. 31)

Of course, what exactly a relational environmental ethic would entail needs to be further fleshed out, but the approach is beginning to coalesce as a genuine possibility within environmental philosophy.

In this regard, it is important to see that it is not just East Asian philosophical traditions that view relationships as central to ethics. Although virtue ethicists have not traditionally talked in terms of relationships, virtue is fundamentally about relating well to others—and can be extended to include nonhuman others. What has, perhaps, hampered the development of this relational dimension is the focus on individual character as something *contained within* the Western autonomous self rather than virtue as a capacity for relating well to others, whose expression always occurs in interaction.[3] Twentieth-century Japanese philosopher Watsuji Tetsurô, who worked to integrate insights from Asian and European philosophical traditions, argued that at the heart of ethics is "betweenness." As Robert Carter explains Watsuji's view:

> We enter the world already within a network of relationships and obligations. Each of us is a nexus of pathways and roads, and our betweenness is already etched by the natural and cultural climate that we inherit and live our lives within. The Japanese live their lives within this relational network. It is imperative, therefore, that one know how to navigate these relational waters successfully, appropriately, and with relative ease and assurance. The study of these relational navigational patterns—between the individual and the family, self and society, as well as one's relationship to the environment—is the study of ethics. (Carter 2009)

For further thought

1. Why is Leopold's *A Sand County Almanac* is a good example of a relational approach to environmental ethics?
2. Do you believe it is possible to achieve Naess's deep ecology ideal of self-realization? Why or why not?
3. Identify one strength and one weakness of applying a Confucian perspective to environmental ethics.
4. How might an emphasis on intrinsic value conflict with a relational approach to environmental ethics? How might an emphasis on intrinsic value complement a relational approach?

Conclusion: Care and meaning—toward a relational perspective in environmental ethics

Relational perspectives from East Asian traditions such as Confucianism and Daoism bear interesting parallels to developments in contemporary feminist ethical theory. In particular, an emphasis on the character and quality of human relations is central to the work of feminist philosophers, such as Nel Noddings (2003), who have developed a theoretical perspective called the *ethics of care*. Noddings' approach takes caring relations as central, and the mother–child relationship as the paradigmatic caring relation. Thus, like classical Confucians, Noddings holds that early family relationships are at the heart of ethics. Although there is some disagreement about the centrality of the mother–child relationship specifically, there is a substantial body of work on care ethics, tied together by the conviction that care and empathy are at the core of ethical life. Although empathy with nonsentient parts of nature may be impossible (insofar as empathy involves mirroring the emotional experience of the individual with whom one empathizes), it is certainly possible to extend the idea of care, and of caring relations, to animals, plants, and ecological systems.

Although an emphasis on care holds significant promise, a relational approach in environmental philosophy need not focus on care, or even on ethics per se. Philosopher Alan Holland (2012) argues that environmental philosophy should move away from a focus on intrinsic value, and even a focus on thinking of our relationship to the natural world primarily in ethical terms. Instead, he suggests that we focus on what he calls "the value-space of meaningful relations" (Holland 2012, p. 3). To focus on meaningful relations emphasizes humans' interaction and engagement with the natural world, and it cuts to the core of what is important in human life. In Holland's view, a focus on intrinsic value tends to obscure critically important environmental considerations. Intrinsic value, as traditionally construed, is detached value—value that exists independent of a particular valuer—whereas relations are fundamentally about attachment and connection. Attachment and connection are important in environmental decisions, because our attachments to particular places and entities matter: they embody meaningful relations. Holland further suggests that meaningful relations are not restricted to relations involving humans. Ecological and evolutionary relationships also carry meaning:

> Natural, in the sense of biospherical, relationships . . . are a paradigm of meaningful relationships both on account of the (past) history invested in them and on account of the (future) history they foreshadow. They encompass, for example, all those biotic relations that make evolution, speciation, and biodiversity possible . . . (Holland 2012, p. 11)

If we reorient toward a relational approach to environmental philosophy, we will focus on relationships as a source of value, and on the quality of relationships as something that deserves careful attention. We can see through a relational perspective how our identities are bound up with others—not just with other human beings, but with nonhuman animals—whether our pets, the animals we eat, or the wildlife in local parks—and with particular places.

A relational approach seems to be one way of moving toward the Leopoldian ideal in which we consider ourselves members of the biotic community, connected and interdependent in diverse ways with myriad living things. The task of ethics, from this perspective, is not to maximize value or conform to an abstract moral law, but to figure out how to engage and relate well. The relational approach, as developed thus far, focuses primarily on relations in a dyadic sense: one person and her relations with individual persons, places, or organisms. In Chapter 4, we turn to the larger sphere of social relations and their ethical significance.

Further reading

Callicott, J. B. (1987). "The conceptual foundations of the land ethic," in J. B. Callicott (ed.), *Companion to A Sand County Almanac: Interpretive and Critical Essays*. Madison, WI: University of Wisconsin Press, pp. 186–217.

Goodpaster, K. (1978). "On being morally considerable." *Journal of Philosophy,* 75(6), 308–325.

Hourdequin, M. and Wong, D. (2005). "A relational approach to environmental ethics." *Journal of Chinese Philosophy,* 32(1), 19–33.

Leopold, A. (1949). *A Sand County Almanac and Sketches Here and There*. New York: Oxford University Press.

McShane, K. (2007). "Why environmental ethics shouldn't give up on intrinsic value." *Environmental Ethics,* 29, 43–61.

Naess, A. (2005). "Self-realization: An ecological approach to being in the world," in A. Drengson (ed.), *The Selected Works of Arne Naess*. Dordrecht, The Netherlands: Springer, pp. 515–530.

Regan, T. (2004). *The Case for Animal Rights*. Berkeley, CA: University of California Press.

Rolston III, H. (1988). *Environmental Ethics: Duties to and Values in the Natural World.* Philadelphia, PA: Temple University Press.

Singer, P. (2002). *Animal Liberation.* New York: HarperCollins.

Stone, C. D. (2010). *Should Trees Have Standing? Law, Morality, and the Environment.* New York: Oxford University Press.

Taylor, P. (1981). "The ethics of respect for nature." *Environmental Ethics,* 3, 197–218.

4

The social dimensions of environmental problems

Introduction

One important diagnosis of our problematic relationship to the natural world focuses on anthropocentrism and our associated failure to value nonhuman organisms and ecosystems. This diagnosis has generated extensionist approaches to environmental ethics, which broaden the circle of moral concern from humans to animals, plants, and beyond. In general, however, extensionist approaches give little attention to relationships *among* human beings, and how these in turn might be connected to environmental problems. This chapter aims to reveal some of the connections between environmental problems and social structures, particularly structures of hierarchy, domination, and injustice.

The first section of the chapter discusses *ecofeminism*, a philosophical approach that emphasizes connections between the domination of women and the domination of nature. Ecofeminism also ties into the discussion of relational approaches to ethics in Chapter 3, as many ecofeminist writers have sought to make visible the importance of contexts, relationships, and narrative in our understanding of and responses to the natural world. Ecofeminists have also argued that certain key dimensions of ethical life—such as care and love—have been backgrounded in traditional moral theories, and they have highlighted the ways in which these elements of ethics can contribute to the development of an environmental ethic.

In the following section, we turn to questions of environmental justice. Theories of environmental justice share some critical concerns in common with ecofeminist approaches. In particular, the environmental justice movement, and the theory behind it, aims to identify and eliminate unfairness in the distribution of environmental burdens and benefits. Theories of environmental justice aim both to clarify what justice requires and identify sources of injustice. Insofar as injustice reflects not only individual prejudice but also social, cultural, and institutional discrimination and oppression, ecofeminism and environmental justice share a common commitment to offering structural critiques of society. By exploring the connections between discriminatory and oppressive conceptual structures and similarly problematic institutional structures, we will also begin to see more clearly the connections between ethics and institutions, a theme at the heart of Chapter 5.

Environmental justice is often construed as *intragenerational justice*: justice within generations, or justice in the distribution of environmental goods and harms among currently living people. However, environmental justice also has a temporal dimension. How can we fairly distribute environmental goods and harms *over time*? This question involves *intergenerational justice*, and the concept of sustainability notwithstanding, theorizing about intergenerational justice remains underdeveloped. This is due, in part, to philosophical problems such as the *nonidentity problem*, which we shall discuss.

The conclusion of the chapter returns to questions and themes from Chapter 3 regarding anthropocentrism and nonanthropocentrism. Can an ecofeminist, justice, or rights framework adequately address the full scope of our environmental concerns? Can any of these approaches provide sufficiently nonanthropocentric grounds for protecting individual living things, species, and ecosystems? Justice frameworks typically focus on human beings, and ecofeminists have been critical of nonanthropocentric approaches such as deep ecology. Thus, there are often tensions between biocentrism and ecocentrism, on the one hand, and concerns of environmental justice on the other. There may be ways to ease these tensions, however, and the chapter closes by considering these possibilities.

Ecofeminism

Ecofeminism draws insights from feminist theory and environmental philosophy, as well as from social theories and movements focused on the elimination of oppression. At the core of the ecofeminist perspective is the idea that the domination of women and the domination of nature are importantly linked (Warren 1996). The term "ecofeminism" was coined by Françoise D'Eaubonne in the 1970s, and the early development of ecofeminism drew on the insights of feminist theorists such as Simone de Beauvoir, who argued that patriarchal systems treat both women and nature as "other" (Glazebrook 2002). The theoretical elaboration of ecofeminist theory has called attention to the feminization of nature and the naturalization of women—the establishment of conceptual associations between women and nature in which both occupy subordinate positions in relation to men and civilization, respectively. Although ecofeminist theorists such as Karen Warren argue that ecofeminism offers insights both to feminism and to environmentalism, the central aim of ecofeminism arguably has been to inject a different perspective into environmentalism, environmental philosophy, and environmental political theory. For example, ecofeminists have criticized environmental philosophers for paying too little attention to the way in which oppressive and exploitative *social* relations are tied to an oppressive and exploitative relationship to nature. The fundamental shared conviction of ecofeminists is that in order to establish a positive, nondominating relationship with nature, we need to address the problematic patterns of domination that exist within society. The many forms of oppression and domination common to contemporary societies—sexism, racism, classism, and so on—are all intimately linked.

If this is the case, then why focus on the oppression of women and the environment in particular? Why not tackle all forms of oppression simultaneously? In principle, ecofeminism *does* support the elimination of all forms of oppression, however ecofeminists see particular reasons to focus on women and the environment, because there are important historical and conceptual connections between these two forms of oppression. We saw in Chapter 1 the ways in which Carolyn Merchant argues for historical connections among women, nature, and environmental exploitation. The conceptual connections, which have been developed especially clearly by philosopher Karen Warren, are explored in the following section.

Dualisms, up/down thinking, and the logics of domination

Especially in the West, we are accustomed to dualistic modes of thinking. "Dualism" may sound sophisticated, but this term describes a familiar way of conceiving opposites as contradictory and mutually exclusive. Cold and hot, up and down, short and tall, dark and light: these are all opposites, and they are typically understood as distinct and as excluding one another. If it is dark, it cannot also be light. If I am short, I cannot at the same time be tall (though I may be short relative to Michael Jordan and tall relative to a three-year-old). There are other ways of thinking about opposites, however, in which they are intimately related. Such views tend to be more common in East Asian traditions of thought, where male and female, for example, are understood to be mutually interdependent. As Thomas Kasulis (2002, p. 87) explains, while the dualistic view of opposites "emphasizes [their] radical separateness," alternative views "[emphasize] the inseparability within their differentiation."

Conceiving of opposites in dualistic and mutually exclusive terms has been a strong theme throughout Western philosophy. This extends to ethics, where value dualisms—in which one member of a binary pair is valued over the other—are common. When we combine a penchant to divide the world into polar opposites with the tendency to value one pole and devalue the other, we have what Karen Warren (1996, p. 21) calls "up-down thinking," which manifests itself as a tendency to understand difference in binary terms and to identify one of the binaries as good (or "up") and the other as bad (or "down").

Think about the history of racism in the United States, for example. A clear binary between "black" and "white" provided the basis for an associated value dualism, and up/down thinking in which "white" was associated with superiority and "black" with inferiority. This in turn played into a racist *logic of domination*: "a structure of argumentation which leads to a justification of subordination" (Warren 1996, p. 21). Such arguments work by identifying binaries, marking one binary as superior over the other, and asserting that this superiority justifies the subordination of the inferior member of the pair.

In the environmental context, we can see how the nature/culture binary sometimes functions in this way. Although the status of "nature" and "culture" as mutually exclusive opposites is now widely contested, this dualistic way of thinking has nonetheless been highly influential. Environmental historian William Cronon (1995) argues that the nature/culture dualism is, in fact, at the heart of the idea of protecting wilderness as land that is pristine and free of human influence, and this point plays an important role in his controversial critique of the wilderness idea. In valuing wilderness, argues Cronon, we are placing nature in the superior position, and consequently devaluing the human and the cultural. This is deeply problematic, as it leaves no space for us to act well in the world. The up/down thinking at the heart of the wilderness idea vilifies humans: nature is pure and good, and humans can only sully and disrupt its perfection. Ironically, this may lead us to neglect nature in the places we inhabit every day, such as cities, towns, and rural areas. Whether or not one agrees with Cronon's analysis of wilderness, his argument does clearly illustrate the way in which up/down thinking can box us in, creating limited space for the negotiation of creative and positive relationships across difference.

From an ecofeminist perspective, the nature/culture binary typically works not to elevate nature over culture, but rather the reverse. What's more, the nature/culture dualism and the female/male dualism are conceptually linked. We see this in the association of men and culture with the mental, and women and nature with the physical (Warren 1996). Combined with the traditional view that rational man is superior to irrational and unthinking nature, the association of women with nature works to justify their occupying a lower rung in the natural order (see Roach [1996] for related discussion).

The idea that men are fundamentally rational and cultural beings, elevated over nature, and women are emotional and biological beings, associated with nature, has not gone unchallenged. Nevertheless, the residues of these associations still remain. Women in the United States, for example, did not win the right to vote until 1920, and the historic exclusion of women from voting, as well as from professions such as the law, relied on the assumption that the natural role of women was as child bearers and nurturers, and that their biology was fundamentally incompatible with any other role. In *Bradwell vs. Illinois*, an 1873 court case in which Myra Bradwell asserted her right to practice law in the state of Illinois, U.S. Supreme Court Justice Bradley wrote:

> [C]ivil law, *as well as nature herself*, has always recognized a wide difference in the respective spheres and destinies of man and woman. Man is, or should be, woman's protector and defender. *The natural and proper timidity and delicacy which belongs to the female sex evidently unfits it for many of the occupations of civil life*. The Constitution of the family organization, which is founded in the divine ordinance as well as in the nature of things, indicates the domestic sphere as that which properly belongs to the domain and functions of womanhood. The harmony . . . of interest and views which belong, or should belong, to the family institution is repugnant to the idea of a woman adopting a distinct and independent career from that of her husband . . . The paramount destiny and mission of woman are to fulfill the noble and benign offices of wife and mother. (emphases added)

Ecofeminists argue that the historical and conceptual connections between women and nature, and the devaluation of women and nature in relation to men and culture, are critical to understanding contemporary environmental problems. Without exposing and overcoming the logics of domination that function with respect to women, nature, racial minorities, and other oppressed groups, we cannot establish a healthy and thoughtful relationship to the natural world.

Varieties of ecofeminism

While ecofeminists generally agree that the oppression of women and nature are importantly linked and should be tackled in concert, ecofeminist philosophy encompasses a wide range of ideas. Thus, ecofeminism is best understood as a theoretical approach rather than as a specific, fixed theory. Karen Warren describes ecofeminism as a patchwork quilt, unified by common borders, yet marked by a diverse array of patterns and colors within those borders. While some ecofeminists focus on the conceptual and philosophical connections between gender and nature, others emphasize their contemporary political and social dimensions, for example. Ecofeminists have engaged in critiques of Western rationality, capitalism, technology, and international development. Others have developed ethical theories that emphasize context, care, narrative, or relationships, and still others have offered ecofeminist defenses of veganism, or elaborated ecofeminist spiritual views. While the full range of ecofeminist theories cannot be covered in a single chapter, some examples can help illustrate both the variety of approaches as well as common themes. Ecofeminist analyses of animals, rationality, and international development will be discussed below.

With respect to animals, ecofeminists have criticized both standard feminist theory and animal rights theories such as Singer's and Regan's, developing alternatives that call us to reconceive our relationships with other sentient beings. Lori Gruen, for example, argues that much of feminist theory—which she labels "anthropocentric feminism"—supports human superiority over animals. Liberal feminism, Marxist feminism, and socialist feminism all come in for criticism. Liberal feminism is problematic because it does little to challenge the patriarchal status quo. Rather than asking that society be fundamentally transformed, liberal feminism seeks an equal place for women within the existing system—a system which, argues Gruen, is fundamentally exploitative of animals and the planet. Marxist feminism, though more radical in its analysis of capitalism and the exploitation of the working class, nevertheless leaves in place a hierarchical relationship between humans and nature. Similarly, socialist feminism, though its critique of traditional institutions is far reaching, nevertheless has "not yet addressed the institutionalized oppression of animals and its relation to oppression generally" (Gruen 1993, p. 77). For traditional feminists, insofar as women's domination is facilitated by the association between women and nature, the solution is to eliminate the association, taking women out of nature and asserting their place in the cultural realm. From an ecofeminist perspective, this is problematic, since it leaves unchallenged the value dualism between culture and nature.

From the perspective of many ecofeminists, animal rights theories such as those of Singer and Regan similarly preserve problematic dualisms (Gruen 1993; Plumwood 1996). The binary opposition of reason and emotion is particularly worrying. In this dichotomy, morality is traditionally placed within the realm of reason, leaving little space for moral emotions such as compassion, which do not fit on either side of the divide (Plumwood 1996, p. 158). This is particularly problematic because situationally sensitive moral emotions are critical to human–animal relationships. The universalist logic of traditional Western moral theory cannot account for the particularity and contextuality of our relationships with other living beings (Plumwood 1996; Kheel 2008) and without an emotional connection to animals, we may fail to be motivated to care about them. Thus, the rationalist arguments of Singer and Regan—while intellectually compelling—may nevertheless leave us cold when the time comes to act. It is not uncommon, for example, for those who encounter Singer's case for vegetarianism to feel the argument's *logical* pull, yet remain unmoved to change their eating habits, which are deeply and intricately tied to emotions.

One response to this problem is to embrace the emotional dimensions of our moral lives, developing a care-centered animal ethic. Worrying that masculinist theories are too quick to focus on wholes such as ecosystems and species at the expense of individuals, ecofeminist Marti Kheel (2008, p. 224) recommends that we refocus on *appropriate care*, which enables responsiveness to particular, situational features. Such an approach, she suggests, supports *contextual vegetarianism* (see Curtin 2004):

> Instead of advocating a universal injunction to be vegetarian based on abstract, atomistic criteria, the contextual approach seeks to understand vegetarianism as a response to particular social and cultural networks of relations. (Kheel 2008, p. 234)

Contextual vegetarianism does not insist on the categorical moral necessity of foregoing meat at all places and times. Instead, it focuses our attention on the relationships in which we embed ourselves when we choose to eat meat. Are these relationships of oppression and domination? What kind of life did this animal have, prior to its slaughter? If I had watched this animal being raised, would I still feel comfortable eating it? In some instances, it may be possible to eat meat without entering into dominating and oppressive relationships with animals, or it might be necessary to eat animals to survive. Context matters, but in the case of relatively wealthy consumers purchasing meat produced by contemporary industrial animal agriculture, it is unlikely that we can honestly describe meat eating as a practice consistent with ecofeminist ideals.

Elements of these arguments regarding animals connect with ecofeminist critiques of the dominant Western conception of reason. Val Plumwood (1996) identifies the reason/emotion dichotomy as a particular concern. It is tied, she argues, to a view that emphasizes the intellect as the critical feature of persons. This rationalist conception of the self, in turn, supports both a strong dualism between humans (as rational beings) and nonhumans (as irrational), as well as a split *within* humans, between our irrational, animal natures and our rational, human natures. Western philosophy generally has endorsed this split and advised us to overcome, tame, or transcend our animal natures in order to become fully human. However, this excludes important aspects of our full selves. In prizing rationality, universality, and impartiality, the traditional conception of reason not only makes no place for the moral emotions, it also devalues the physical realm and the significance of our bodily existence; generalizes in ways that overlook individual differences; and diminishes the importance of care (Plumwood 1996).

The impetus for universality, a powerful Enlightenment ideal, comes in part from the perceived dangers of special relationships and parochial concern. We are, of course, all familiar with the parent whose own child can do no wrong: it is always *other* children who are the instigators of mischief. It is also true that corrupt bureaucracies around the world thrive on special relationships and special favors. In many cases, the only way to get treated well is to know (and perhaps bribe) the right people. Yet recognizing the dangers of special relationships or particular affections should not lead us to conclude that special relationships be abolished altogether in favor of contextually insensitive, equal treatment for all. Surely the best teachers, coaches, and parents, for example, are those who recognize the unique needs of those whom they nurture, and tailor their guidance accordingly. When Plumwood critiques dominant conceptions of rationality, she is not arguing that we should do away with impartiality altogether: there are contexts where impartiality is important and necessary. What Plumwood tries to do, however, is to break down the strong separation created by reason/emotion and human/animal dualisms.

In breaking down these dualisms, we find ourselves with a wider array of possibilities for conceiving reason, emotion, and their relationship. Dualistic thinking limits our options. Plumwood argues that much of environmental philosophy, though it offers important critiques of anthropocentrism and instrumental value, remains constrained by conventional concepts and categories. Her suggestion, then, is that we should not value the devalued by inverting our up/down thinking—elevating emotion, for example, above reason. Instead, we should question dualistic thinking itself, with its inherent oppositions:

> Part of a strategy for challenging this human/nature dualism . . . would involve recognition of . . . excluded qualities . . . as equally and fully human. This would provide a basis for recognition of continuities with the natural world. Thus reproductivity, sensuality, emotionality would be taken to be as fully and authentically human qualities as the capacity for abstract planning and calculation. (Plumwood 1996, p. 169)

Once we stop thinking of the human and the natural, or emotion and reason, as mutually exclusive categories, new possibilities emerge. Reason, for example, might inform and be informed by emotion. Animals might not be fully excluded from the realm of reason, and rather than being seen as passive, raw material (awaiting our use), the natural world might be seen as active and responsive. Persons might be seen neither as completely apart from nor fully identical with nature, but rather in ongoing, interactive relationship with it.

Having introduced some ecofeminist perspectives on animals and Western rationality, it is now time to turn to questions of international development. Here, we will focus on the work of Vandana Shiva, a scientist, author, activist, and ecofeminist. Like most ecofeminists, Shiva is struck by and deeply concerned about what Warren (1996) calls "the twin dominations of women and nature." In less-developed countries such as her home nation India, Shiva notes that women and nature are closely linked, such that environmental degradation directly and dramatically impacts women's lives and livelihoods. What's worse, "development" often exacerbates the plight of women in many countries by disrupting existing economies and imposing a Western economic model in which women's nonmarket labor is devalued, since it remains invisible to market metrics such as gross domestic product (GDP) (Shiva 1989). What's more, standard Western development models understand "productivity" from a narrow and market-centered perspective:

> A stable and clean river is not a productive resource in this view: it needs to be "developed" with dams in order to become so. Women, sharing the river as a commons to satisfy the water needs of their families and society are not involved in productive labor: when substituted by the engineering man, water management and water use become productive activities. (Shiva 1998, p. 272)

The use of traditional economic models—which typically exclude nonmarket labor and fail to count natural resources as assets unless they are bought and sold, or given market prices—also distorts our understanding of poverty, argues Shiva. In a subsistence economy, standard metrics may fail to capture aspects of wealth and productivity that fall outside of the market. People may produce crops for their own consumption, for example, rather than cash crops for export, or construct their own housing from local materials rather than purchasing wood and other supplies. Such individuals may be able to comfortably meet their basic needs, and seeking more money by participating in market economies (e.g., growing food to sell to other countries) can sometimes lead to greater impoverishment, by displacing subsistence crops and generating dependence on fluctuating market prices.

Shiva (1989, 1998) analyzes Western-style development as associated with increased deprivation for women, who often lose access to water, agricultural land, or forests, as these are privatized and converted to market uses.

Moreover, argues Shiva, development involves the "death of the feminine principle," because it disrupts male–female complementarity, introduces new sources of inequality, and imposes a reductionist world view in which non-Western people, women, and nature are all seen as lacking and in need of improvement (Shiva 1998, p. 273). This worldview similarly promotes Western technology—such as genetically modified seeds, which are tightly bound into large-scale corporate capitalism—over local knowledge and traditional agricultural practices such as seed saving. The work of Shiva and others such as Maria Mies, a German social scientist, introduces additional dimensions to ecofeminist analyses, emphasizing the developed/underdeveloped dualism as deserving critical scrutiny, and calling our attention to the role of markets and technology in imposing narrow and confining modes of development throughout the world (Mies and Shiva 1993).

Objections and concerns

Ecofeminism clearly provides a perspective on our relationship to the natural world that various other approaches overlook. Nevertheless, ecofeminist approaches have generated criticism, including criticism from other feminists. With respect to Shiva's work in particular, we encounter worries about gender essentialism, the idea that women naturally and essentially have certain characteristics, such as nurturing dispositions or an intimate connection with nature (Davion 1998). The language Shiva uses does sometimes suggest a fundamental and essential tie between women and nature, and in doing so, may reinforce gender categories rather than question them (Davion 1998). In particular, by celebrating male–female complementarity, Shiva seems to leave the male/female dichotomy untouched. As Victoria Davion (1998, p. 282) explains,

> To simply accept gender complementarity without exploring questions it raises is to ignore feminist literature claiming the gender roles are part of the means of domination and subordination in patriarchy.

Davion labels approaches like Shiva's, which retain gender categories while elevating the feminine, *ecofeminine* rather than *ecofeminist*. Views suggesting that women are naturally intuitive or in harmony with the earth similarly count as ecofeminine. Rather than challenging standard categories, such views retain them, while reversing the up/down polarity so that the feminine comes out on top, with the masculine being demoted to the lower position. Arguably, this contradicts what Karen Warren and others see as the fundamental spirit of ecofeminism, which involves *challenging* value dualisms rather than merely reversing them. As discussed earlier, ecofeminism can prompt us to revalue the devalued, or to foreground values, like context-sensitive care and compassion, that occupy the background of traditional moral theories such as Kantian ethics. It is not necessary to insist on an inversion of values, so that traditionally feminine values are prized over and above traditionally masculine ones. Loosening the grip of value dualisms makes space for a wider range of ethical values and approaches.

There is another worry, however. The challenge of gender essentialism is closely related to a concern about the connections ecofeminists assert between the domination of women and the domination of nature. Are *all* ecofeminist views guilty of gender essentialism, in virtue of the connection they identify between these two forms of oppression? To this, we can safely answer no: ecofeminists do not assert an *essential* connection between women and nature, or their domination. Instead, they show how our concepts tie together women and nature, and point to historic and contemporary connections between the oppression of both. That is why Karen Warren (1996, p. 35) argues that any adequate environmental ethic must also be feminist: "[W]ithout the addition of the word *feminist*, one presents environmental ethics as if it had no bias, including male gender bias, which is just what ecofeminists deny."

Nevertheless, ecofeminists accept, and indeed take heart in the fact that conceptual connections are not fixed, nor are systems of oppression. Men are often associated with nature, for example—sometimes in contexts that glorify them (witness the rugged toughness of the Marlboro man) and also in contexts that denigrate them (as when a brutal killer is labeled an "animal" or a "wild beast"). Ecofeminist analyses are intended to open up and make visible the ways in which gender and nature are connected, and how these connections are tied to valuations of each.

Despite these clarifications, some still worry that ecofeminists focus too strongly on the ties between gender and nature, suggesting at times that these two forms of oppression are inseparable, and that each can be addressed only in light of the other. However, it does seem that an environmental philosophy may truly stand against oppression of nature—as does deep ecology—without necessarily addressing the oppression of women (Glazebrook 2002). (Many ecofeminists have, in fact, criticized deep ecology on just these grounds.) The oppression of one group or entity may be overcome even if oppression of others remains. The varying rates of progress on different struggles for recognition and fair treatment—such as those focused on women, animals, racial minorities, homosexuality, or minority religious groups—provide support for this view. Thus, while ecofeminism offers an important and valuable perspective in environmental philosophy, it is not clear that an ecofeminist approach is the only legitimate one. Employing a familiar metaphor, perhaps we should think of ecofeminism as having a broader place in a diverse quilt of environmental thought. The ecofeminist aspiration to overcome oppression of all kinds is an important and crucial one, but at times we may be forgiven if we focus on a particular problem, or even a particular aspect of it, without addressing all elements of injustice at once.

Theoretical and practical significance of ecofeminism

Regardless of the concerns just discussed, the ecofeminist perspective has brought important new dimensions to discussions in environmental philosophy. Three of ecofeminism's most important contributions include the following: (a) highlighting the confining nature of our conceptual frameworks and opening up new possibilities for ethics and environmental philosophy; (b) drawing out these possibilities through the development of the role of narrative and context in ethics, a more relational understanding of the self, and an ecofeminist care ethics; and (c) drawing connections between theory and practice through ecofeminist activism. We have already discussed in some detail the first point, so it is to the latter contributions we now turn.

Although ecofeminist scholarship has had a strong critical orientation, analyzing and diagnosing inadequacies in standard approaches to environmental ethics, this critical work lays the foundations for the development of more substantive, positive theory. For example, in the second half of her article, "The power and promise of ecological feminism," Karen Warren discusses in detail the ways in which ecofeminist thought may help us reorient our approaches to ethics generally, and to environmental ethics more specifically. Here, Warren highlights the relevance of care, relationships, narrative, and context to environmental ethical thinking. Through narrative—specific stories recounting our experiences in nature—we see the role of care, relationships, and context. Our interest in and care for the natural world are often motivated and triggered by specific encounters: a squirrel climbing the stalk of a sunflower to reach the seeds, a fox crossing a field at sunset, or a tide pool teeming with life. It is in these encounters that we notice the texture and diversity of the biotic and abiotic world. Warren (1996, pp. 27–28) argues that there are four reasons to take seriously the role of narrative in feminist environmental ethics: first, narrative brings forward "a felt sensitivity often lacking in traditional analytical ethical discourse . . . It is a modality which *takes relationships themselves seriously*"; second, narrative "gives expression to a variety of ethical attitudes and behaviors often overlooked . . . in mainstream Western ethics"; third, it emphasizes ethics as involving insight that emerges from specific situations rather than as rules imposed upon them; and finally, narratives can act as arguments, since the arc of a narrative provides a sense of what comes next, and what a fitting response to a particular situation might be.

Warren's account suggests that narrative, context, and relationship are all importantly interrelated in ethics. Others echo and further develop these ideas. In keeping with our discussion of relational approaches to ethics in Chapter 3, Val Plumwood (1996, p. 172) argues that a relational account of the self can "clearly [recognize] the distinctness of nature but also our relationship and continuity with it." Plumwood worries that some versions of deep ecology recommend an expansive self that encompasses the entire natural world, and that this is a dangerous "extension of egoism" (p. 166). A relational approach avoids this pitfall. Rather than assimilating nature into individual human selves, a relational approach conceives the "self as embedded in a network of relationships with distinct others" (p. 172). This avoids both the atomism and individualism of traditional moral theory and what Plumwood sees as the egoistic expansiveness of some deep ecological views. By emphasizing interconnection and the role of others (including nonhuman others) in constituting our identities, the relational approach helps thwart a narrowly instrumentalist outlook on nature, where animals, plants, and the natural world are seen as mere resources for our use.

Of course, acknowledging relationships and interdependency is not itself sufficient to guarantee a *positive* relationship with the natural world. As noted above, there are damaging and exploitative relationships, and clearly these are not the basis for an ethical connection to nature. Thus, as noted in Chapter 3, the quality and character of our relationships is an important ethical focal point. As Chris Cuomo (1998, p. 98) explains, "Though they emphasize the social nature of human individuality, feminists remain critical of the ways in which social relations and formations are damaging—producing and reproducing prejudice as well as connectedness, oppression as well as resistance, confining norms as well as identity." The development of care ethics, as well as its application to the animals and the environment, subjects relationships to critical scrutiny, and emphasizes context-sensitive, appropriate care as central to positive relations, as discussed above.

In addition to opening new conceptual spaces and developing ethical outlooks that both complement and challenge mainstream moral theories, ecofeminism has had strong ties to activism and social movements against oppression. This connection may, in part, emerge out of women's standpoints and experiences of discrimination and sexism. One of the key ideas of *feminist standpoint theory* is that members of marginalized groups may see aspects of a situation that are invisible to others (Harding 1991). Such individuals may ask different questions in scientific research, for example, questioning standard assumptions. In the environmental realm, women's experiences of oppression may sensitize them to the exploitation of nonhuman animals and the natural world. A vivid and extreme example of this is described by ecofeminist author Chaone Mallory (2006), who recounts the story of a female forest activist who is raped by a fellow activist, who is male and has been assigned to teach her how to defend forests by occupying trees slated for cutting. The female activist not only describes her own feeling of vulnerability and violation, but ties this to the position of the forest and the logging practices that her activism opposes, finding an important connection between assertion of power over another person and the domination and destruction of the forest.

Following in the footsteps of the feminist tradition more broadly, ecofeminist theorists have self-consciously considered the relationship between theory and practice, seeking ties between the academic and political realms. Ecofeminists have reached out to the public, for example, in developing arguments for vegetarianism and veganism (see, e.g., Adams 1990). Although they do not explicitly endorse the tenets of ecofeminism, many important international organizations—including the International Union for the Conservation of Nature (IUCN) and the Organisation for Economic Cooperation and Development (OECD)—recognize key connections between gender and the environment. Discussions of gender and the environment in these international contexts are frequently tied to issues of sustainable development and environmental justice. The next section introduces environmental justice, both within and across generations.

For further thought

1. What are the similarities that exist between the oppression of women and the oppression of nature?
2. Why might the nature/culture binary be problematic? Are there ways in which the nature/culture binary can benefit the environment?
3. What are some advantages of a contextual and care-centered ethic, as opposed to traditional environmental and animal rights ethics? What are some disadvantages?
4. What distinction does Victoria Davion draw between "ecofeminine" and "ecofeminist" views? Why is an "ecofeminine" ethic potentially problematic, in her view?
5. What, in your view, is ecofeminism's most important contribution to environmental ethics? Why?

Environmental justice

Concerns about environmental justice play a key role in contemporary environmental thought and politics. It has become increasingly obvious, for example, that certain groups—women, racial and ethnic minorities, the poor, and the politically disempowered—bear undue burdens associated with pollution and other environmental problems. In the United States, Native Americans have borne heavy burdens in association with uranium mining in the western states, and Latino farm workers are disproportionately exposed to toxic pesticides (Shrader-Frechette 2002, p. 9). At the international scale, some of the strongest impacts of global climate change are being felt by the world's poorest people. In the meantime wealthy countries continue to exploit the resources of less-developed nations and burden these same nations with toxic wastes generated in the United States and Europe (Pellow 2007).

In some cases, disparate environmental burdens are defended as a natural outcome of the free market: housing in polluted areas is less expensive than in neighborhoods with clean air and water, while housing near environmental amenities such as parks and open space costs more. However, in the United States at least, economic disparities cannot fully account for the greater burden experienced by African Americans and other minorities (Bullard et al. 2008). What's more, the economic logic itself seems to beg the question of environmental justice. Should people have to breathe dirty air and drink polluted water, merely because they are poor? What does justice require?

Justice in the present

Concerns about the distribution of environmental harms or "bads" such as pollution among current people have been front and center in the environmental justice movement, as well as in theorizing about environmental justice. Nevertheless, environmental justice encompasses a broader range of concerns. For example, it is not just the ultimate distribution of pollution that is a matter of justice, but also the processes used to decide how to manage pollution. In addition, although many theories of justice focus on existing persons, there are also important questions of intergenerational justice, or justice to future generations. These questions are especially pressing in light of global climate change, whose effects lag behind the emissions of greenhouse gases and will extend centuries into the future. Other environmental challenges—such as the disposal of nuclear waste—involve time scales as long or longer. The following section turns to questions of intergenerational justice. For now, we will focus on justice among existing persons.

Three cases of environmental injustice

The roots of the American environmental justice movement extend back to the early 1980s in Warren County, North Carolina. In 1982, the citizens of Warren County rallied together to protest the siting of a polychlorinated biphenyl (PCB) dump in their community. Although more than five hundred people were arrested for protesting this toxic waste dump, the state of North Carolina proceeded with the PCB facility. In addition, the United States Environmental Protection Agency permitted disposal of the waste extremely close to the water table, in a significant departure from the standard practice (Shrader-Frechette 2002, p. 8). Warren County was one of the poorest communities in North Carolina, with a majority African American population. Although the community lost in their struggle to keep out PCBs, their efforts catalyzed a major study of the relationship between toxic waste and race. In 1987, the United Church of Christ Commission for Racial Justice released a groundbreaking report showing that hazardous waste sites tend to be located in areas with large minority populations (Commission for Racial Justice 1987). Race—even more than income—continues to be a strong predictor of proximity to toxic waste sites in the United States (Bullard et al. 2008, p. 373).

Environmental injustice affects other minority communities as well. Native Americans, who historically were displaced from their lands and forced onto reservations, continue to suffer injustice—including environmental injustice—in their relations with the United States. Native American tribes are diverse, and the specific forms of injustice they face vary depending on social, cultural, economic, and environmental contexts. One of the most egregious forms of environmental injustice involves the nuclear industry in the American West. For example, Navajo lands in Arizona have been seriously contaminated by uranium mining, and beginning in the 1950s, Navajo mine workers were exposed to radiation levels that exceeded the allowable limit by a factor of ninety (Shrader-Frechette 2002, p. 117). Many workers at this site later became ill and died. Moreover, in disposing nuclear waste, Native American lands have been heavily targeted. The U.S. Waste Isolation Pilot Plant (WIPP) in southeastern New Mexico is located close to the Mescalero Apache Reservation, and until the project was cancelled in 2012, the Yucca Mountain high-level nuclear waste disposal site threatened Shoshone and Paiute sacred lands. Although the Yucca Mountain project did not go forward, the public engagement process arguably failed to do justice to Native Americans' concerns. In particular, by emphasizing technical issues and employing a cost–benefit framework, the government process failed to fully comprehend the concerns expressed by local tribes, which focused on the historical importance and sacred quality of Yucca Mountain (Endres 2012). The environmental injustice issues described here thus involve not only undue nuclear burdens, but failures in the participatory process and failures of recognition, two aspects of environmental justice discussed further below.

A final example emphasizes the international dimensions of environmental justice. Although the environmental justice movement is generally seen as originating in the United States, concerns about environmental justice are now central to many activist movements around the world. Environmental justice intersects with social justice, civil rights, occupational health, and environmentalism, and the kinds of issues discussed above, involving pollution and toxic waste, are recapitulated in nations around the world. Often these issues span national borders, as when the use of consumer electronics in developed nations results in the exportation of toxic "e-waste"—electronic waste full of heavy metals and other hazardous compounds—to China, India, and Africa, where the poor dismantle computers, televisions, and other equipment in order to salvage and sell the valuable components for cash. Frequently, children engage in these efforts, burning plastic insulation off of wires in order to access the copper inside, or dipping electronics into acid baths to recover gold from circuit boards (Carroll 2008). These crude recycling processes expose those who engage in them to toxic flame retardants, lead, and numerous other dangerous chemicals. Lax regulations and enforcement allow more than 90 percent of e-waste to "[end] up in dumps that observe no environmental standards, where shredders, open fires, acid baths and broilers are used to recover gold, silver, copper and other valuable metals while spewing toxic fumes and runoff into the skies and rivers" (Bodeen 2007).

These examples illustrate not only laxity in environmental regulation or failure to enforce existing standards, but unequal protection, where vulnerable and marginalized groups—children, racial and ethnic minorities, or the poor—experience the greatest burdens of pollution and environmental risk.

Theories of environmental justice

Many theories of justice emphasize fairness in the distribution of various goods. These goods may include wealth and particular material goods, as well as rights, rewards, and opportunities. In discussions of environmental justice, the distribution of pollution and other environmental harms has been a key focal point. The distribution of environmental benefits—amenities such as parks, public transportation, and walkable communities—has recently gained attention as well. Yet, justice is not just about whether distributions are fair. We can distinguish between two kinds of justice: substantive and procedural. *Substantive justice* is about what constitutes a fair outcome. A general principle of substantive justice, for example, might be that each person deserves equal access to basic health care services. In the environmental realm, a principle of substantive justice might hold that everyone deserves some base level of environmental quality, thus no person should have to suffer pollution beyond a certain threshold. *Procedural justice*, in contrast, is concerned with the processes needed to ensure fair decisions. A principle of procedural justice might guarantee each person an opportunity to participate in government decisions likely to significantly affect his or her welfare.

The distinction between substantive and procedural requirements is illustrated in environmental law as well as in ethics. For example, the United States has both substantive and procedural environmental laws. The Clean Air Act of 1970 is a substantive law insofar as it sets limits on the emissions of specific pollutants. The National Environmental Policy Act (NEPA) of 1969, however, is primarily a procedural law. Rather than setting particular standards for environmental quality, this law focuses on establishing a sound process for environmental decision-making. NEPA requires that the government study a variety of alternatives for any federal action with significant environmental impacts, produce an environmental impact statement (EIS) considering these alternatives and their effects, and make the EIS available to the public for comment. Of course, a given law—whether procedural or substantive—may be just or unjust, and a single law alone is unlikely to be sufficient to ensure environmental justice in a nation, state, or municipality. Nevertheless, concerns about environmental justice prompted U.S. President Bill Clinton in 1994 to issue Executive Order 12898, which requires each federal agency to

> make achieving environmental justice part of its mission by identifying and addressing, as appropriate, disproportionately high and adverse human health or environmental effects of its programs, policies, and activities on minority populations and low-income populations in the United States and its territories and possessions . . .

This, in turn, generated new guidelines for taking environmental justice into account during the NEPA process (Council on Environmental Quality [CEQ] 1997). Concerns about procedural environmental justice call attention to the need not only to provide opportunities for public participation, but also to provide opportunities that genuinely enable engagement by all. Among other efforts, this may require steps to "overcome linguistic, cultural, institutional, geographic, and other barriers to meaningful participation" (CEQ 1997).

Some hope that substantive justice can be achieved through procedural justice alone. Procedural accounts of justice—such as that of twentieth century political philosopher Robert Nozick (1974)—hold that a distribution is just if and only if it is the result of a fair process. But advocates of environmental justice tend to focus on both process and outcomes. In practice, having substantive goals for outcomes as well as seeking a fair process can mutually support one another. If a process systematically excludes some groups from participation, for example, there is reason to suspect that the substantive outcomes may be unfair. Conversely, if the substantive results of environmental decision-making consistently burden some groups more than others, the decision-making process deserves scrutiny.

Although there is no consensus as to what constitutes environmental justice, from an ethical perspective, the current state of affairs is difficult to defend. In the United States, membership in a minority group increases substantially one's likelihood of exposure to environmental pollutants as well as pollution-related illnesses (Shrader-Frechette 2002). Around the world, it is clear that climate change—though generated largely by the emissions of industrialized nations in the global North—will place the heaviest burden on those in poor and less-industrialized nations of the global South.

There are many possible substantive principles of justice one might offer to explain what is wrong with the current situation and chart a path forward. One very simple principle would hold that environmental burdens and benefits should be distributed equally. However, achieving such a distribution would require homogenizing the world so that each person experienced identical—or at least equivalent—environments. In addition, the principle seems overly ambitious from a theoretical perspective. We allow for inequalities in many realms, and some degree of difference in wealth, status, job benefits, and so on is not generally considered morally problematic. Nevertheless, it seems perverse that some people should be deprived of environmental quality due to a morally irrelevant factor such as race, and that those who benefit the most from polluting should be least burdened by that pollution.

Philosopher Peter Wenz (2012) builds on this last insight in his proposal, arguing that the distribution of environmental burdens should conform to the Principle of Commensurate Burdens and Benefits, which suggests that each of us should take responsibility for the harms generated by our actions. The basic idea of the principle is that many of the benefits of contemporary society—cars, houses, central heating, air conditioning, computers, and flat screen TVs, for example—come with associated environmental burdens. The mining of metals for semiconductors and the disposal of electronic waste, the drilling for oil and digging for coal used to heat homes, and the greenhouse gases produced by the burning of fossil fuels all damage and pollute the environment. Wenz's idea is that those who consume the most, in general, generate the most pollution, thus it is they who should bear the greatest burdens. Clearly, this idea is unlikely to go over very well with those who enjoy jetting around the world on greenhouse gas intensive trips, or who own capacious mansions with heating requirements to match their size. In Wenz's proposal, the biggest consumers (typically, he argues, the wealthy) should breathe the dirtiest air, drink the dirtiest water, and live closest to toxic waste dumps, landfills, and other locally undesirable land uses (LULUs).

Despite its eyebrow-raising quality, many are hard pressed to say why Wenz's solution would be *unfair*. Impractical, perhaps, given the distribution of political power and the relationship between wealth and power, but unfair? It is a proposal worth considering, at least. If Wenz is wrong about what would be fair, then what alternative principle would be better?

The impracticality objection to Wenz's proposal brings us back to questions of procedural justice. If one thinks that the Principle of Commensurate Burdens and Benefits is morally sound, yet that the principle would never work in practice, why might this be? Arguably, because the wealthy and powerful would never accept a system in which they had to bear the greatest environmental burdens. However, decision-making processes in which the preferences of the wealthy and powerful inevitably trump those of the less well off are arguably not fair, and certainly seem to fit poorly with the basic ideals of democracy. Thus, as suggested earlier, there is an important connection between substantive and procedural justice, since substantive justice (whatever that might entail) is unlikely to be achieved in the absence of fair procedures. Although fair procedures may not guarantee a just outcome, they certainly can help.

Procedural justice gets complicated, though, when one considers that having an opportunity to participate—to express one's views in a public hearing, for example—is not the same as being heard or recognized as having a legitimate role in the decision-making process. In recent decades, theorists such as Iris Marion Young, Nancy Fraser, Axel Honneth, and Charles Taylor have argued that *recognition* is critical to justice. Take a case recounted by Celene Krauss (1994, cited in Schlosberg 2007, p. 61), in which officials address all the white women at a public hearing by their last names such as "Mrs. Smith," but when a black woman comes to testify, she is called by her first name, "Cora." Although being addressed by one's first name may not be insulting or degrading in itself, in this context, the different form of address marks a clear difference in status. Not all participants in the hearing were recognized as equals whose perspectives deserved similar consideration.

Lack of recognition is a problem at many levels. It affects both individuals and social groups (Schlosberg 2007) and encompasses patterns of cultural domination and oppression, invisibility, and disrespect (Schlosberg 2007, citing Fraser 1998, p. 7). Often, lack of recognition results from a failure to understand and appreciate differences, as exemplified in the Yucca Mountain case described earlier. Another example (Schlosberg 2007, p. 60, drawing on LaDuke 2002, p. 60) further illustrates this point. EPA standards for the pollutant dioxin in fish are based on average fish consumption in the United States. However, Native Americans who engage in subsistence fishing practices consume vastly more than this average. Thus, standards that might be protective for some groups do not adequately protect others, whose distinct cultural practices are not fully recognized or incorporated into the standard setting process. Similar issues arise when pollution standards use adult men as the model, failing to recognize differential sensitivities of women and children.

Of course, problems of recognition are not unique to the United States. The international network Peoples Global Action has argued that World Trade Association policies overlook cultural differences and fail to acknowledge diversity, instead imposing a single narrow market economic model around the world (Schlosberg 2007, p. 86). Although some theorists are inclined to collapse recognition into the general category of procedural justice, others argue that treating it as a distinct aspect of justice helps make visible the forms of oppression that undermine the possibility of equal participation and procedural equity (Schlosberg 2007, p. 26).

The idea that lack of recognition is disabling, and that it thereby undermines the possibility of full participation, connects to another approach to justice. This approach, developed by economist Amartya Sen and philosopher Martha Nussbaum, emphasizes justice as supporting the development of key human capabilities. For Nussbaum (2000, pp. 78–80), these capabilities include—among others—life, bodily health, and "[b]eing able to live with concern for and in relation to animals, plants, and the world of nature." Nussbaum (2000, p. 6) argues that a just society should provide a certain "threshold level of each capability" for each of its citizens.

Clearly, life and bodily health require a certain minimal level of environmental quality. Nussbaum's capabilities approach, therefore, seems to demand that no person be burdened beyond this level, and that states have a responsibility to ensure this.

David Schlosberg (2007) argues that the capabilities approach is relevant not only to individuals, but also to communities. Environmental justice is importantly tied to community functioning, and the flourishing of communities depends on the sustenance of certain community-level capabilities. For example, globalization of the food system has led—somewhat ironically—to greater food insecurity for many in developing nations (Schlosberg 2007, Shiva 2000). Through the modernization, industrialization, and marketization of small-scale farms in nations such as India, farmers develop greater dependency on global markets and multinational agricultural corporations (Shiva 2000). They shift from producing food for their own consumption to growing food for external markets, and become subject to the vagaries of market fluctuations in crop prices. This undermines not only individual capabilities, but community capabilities associated with the provision of food and the maintenance of important cultural practices and food traditions, such as saving seed and maintaining diverse varieties of seeds. A capabilities perspective thus highlights the injustice of practices that undermine the functioning and flourishing of individuals and communities, and more positively, focuses on the importance of providing the conditions that support key capabilities for individuals and communities worldwide.

Theories of environmental justice take a variety of forms and emphasize different aspects of justice. It is generally recognized that we need to attend to both procedural and substantive justice, and that the two are importantly linked. A distinct focus on recognition has helped reveal the institutional and structural obstacles to participation, and to problems of "uptake." While decision-makers may make active efforts to include marginalized groups and individuals, if these groups and individuals are viewed as having lower status, or if they express concerns that don't fit the background frameworks and assumptions of the decision-making process, their perspectives may not be fully received and considered. There are also structural features that may make certain individuals and groups more hesitant to speak out, for fear that they will be ridiculed or subject to retaliation. This might be the case, for example, where workers are pushing for fairer standards from their employers or holding powerful institutions accountable for their actions. In the 1960s, for example, after biologist Rachel Carson released her book *Silent Spring*, detailing the hazards of widespread pesticide use, chemical companies engaged in a sustained campaign to mock and discredit her work. These attacks drew on gender stereotypes to portray Carson as a "hysterical" woman while at the same time criticizing her "unfeminine" intrusion into science (see Killingsworth and Palmer 1995; Smith 2001; Corbett 2001). These approaches took advantage of Carson's position of social disadvantage to attempt to silence and disempower her. It is here that we see both the significance of recognition and the potential power of the capabilities approach to justice in revealing the necessity of developing institutions, discourses, and social conditions that support justice. Merely providing opportunities for participation is not enough.

Justice for the future

The discussion above suggests that procedural and substantive justice are importantly linked, and that procedural justice itself can only be achieved in social and institutional contexts that enable participation and provide full recognition for all. But what about those who are unable to participate altogether? Can we have obligations of justice to future generations, to persons who do not yet exist? Intuitively—without worrying about what philosophical theories say—the answer seems to be yes. As Joel Feinberg (2013, pp. 378–379) very straightforwardly puts it:

> We have it in our power now to make the world a much less pleasant place for our descendants than the world we inherited from our ancestors. We

> can continue to proliferate in ever greater numbers, using up fertile soil at an even greater rate, dumping our wastes into rivers, lakes, and oceans, cutting down our forests, and polluting the atmosphere with noxious gases. All thoughtful people agree that we ought not to do these things . . . Surely we owe it to future generations to pass on a world that is not a used up garbage heap.

However, philosophers have long been skeptical about obligations to future generations, and our current practices do not seem to affirm the conviction that we have a duty to leave the world in good condition for those that follow us many generations hence. This point about practices is underscored by our collective unwillingness to take serious steps to mitigate global climate change, whose impacts will be felt most strongly not next year, or next decade, but in the decades and centuries to follow. Of course, we may have firm and defensible ethical convictions that our practices fail to match. So, perhaps the problem is one of motivation: although we acknowledge obligations to future generations, these obligations simply lack enough pull to engage us. Just as the procrastinator knows that she really *should* start her term paper today, but finds so many more interesting things to do, we know that our choices *should* take account of future generations, though when push comes to shove, we are more captivated by our own near-term interests than by the welfare of those who will follow us decades or centuries later. This psychological explanation cannot directly account for the history of philosophical resistance to obligations to posterity, however. Many have argued that in principle future generations lack rights, and we can have no duties to them. We will review some of these objections below.

One source of objection derives from a social contract view of morality. This view—in addition to those discussed in Chapter 2—has played an important role in Western ethical theory, and although its roots arguably go back to Socrates, the seventeenth century English philosopher Thomas Hobbes is generally credited as the first to develop a full social contract theory of morality. Hobbes asks us to imagine humans living in a "state of nature" where no rules govern, where human social life is full of conflict, and life is "nasty, brutish, and short." In such a situation, argues Hobbes, rational persons would be willing to agree to be governed by a set of basic rules in exchange for the security that general adherence to such rules can provide. To ensure adherence to the rules, the parties to this agreement would put in place a sovereign ruler empowered to enforce the social contract. Morality, according to Hobbes' view, is the set of rules to which rational persons in a state of nature would agree, in order to escape the vagaries of a lawless state.

Key to social contract theory is the idea of reciprocity. I agree not to steal from you, so long as you agree not to steal from me. Moral obligations are owed to those with whom we live, who are members of our moral community in virtue of our shared commitment to the social contract. Although reciprocity is arguably an important aspect of morality, a theory in which all obligations must be reciprocal has clear limitations with respect to animals, the environment, and future generations. All and only those who are party to the social contract are bound by moral obligations, and all and only those who are party to the social contract are owed moral duties. But animals, plants, ecosystems, and future generations are not party to the social contract. So we cannot, on this view, be bound by duties to them. One serious objection to the idea of obligations to future generations derives from this kind of reasoning. If obligations require reciprocity, then there can be no obligations to future people, because they can do nothing for us.

Yet, not all obligations must be grounded in reciprocity. As Daniel Callahan (1971, cited in Shrader-Frechette 2002, p. 102) points out, parents have obligations to their children independent of whether their children reciprocate with gratitude or assistance. Similarly, we have obligations to those who are seriously or terminally ill and cannot reciprocate our care. So, lack of reciprocity is not a decisive objection to intergenerational justice.

A second objection focuses on a potential conceptual problem with obligations to future generations. Future people do not yet exist—and how can we have obligations to something nonexistent? Here, Lucas Meyer (2010) offers a reply. It is helpful to remember that we can be confident that there will be people in the future—many of whom do not exist now—who will have interests and rights. Moreover, our actions in the present can affect those future interests and rights, and insofar as our actions "seriously frustrate the interests of future people, we can violate their future rights" (Meyer 2010). This is true even though future people do not yet exist, and even though we cannot know precisely whom those people will be.

A third, and perhaps the biggest, conceptual obstacle to theories of intergenerational justice and the ethics of future generations derives from the *nonidentity problem* made famous among philosophers by Derek Parfit (1984). Parfit argues that understanding the ethical dimensions of our relationship to future people is complicated by the fact that many of the policies we undertake today will influence whom those future people will be. The very identities of the people who are born several generations hence will itself be shaped by the choices we make today. If we choose to conserve natural resources, for example, the world will be a different place 30 years from now than it will be if we choose a policy of wanton destruction and consumption. The differences between these two worlds, in turn, translate into differences in which people meet and interact, who falls in love with whom, and what children are born.

We need not extend too far into the future along these divergent trajectories to find *entirely different individuals* populating the world. Those who will live in the more environmentally degraded world, it turns out, never would have lived had it not been for our habits of destruction and consumption. Given this fact, it seems that those people have no grounds for complaining that *they* have been harmed by our reckless ways. Without our reckless ways, they would not have existed: or put otherwise, there is no possible world in which a particular person in this depleted world could have existed in the absence of depletion. So no person in this world can legitimately say, "I would have been better off if my grandparents' generation had taken better care of the environment."

Many philosophers have concluded on the basis of such arguments that we lack obligations to those in the further future, because no one who exists many generations down the line can claim to be harmed by the actions we take today. However, although the identities of future people are indeterminate and depend on our current choices, this does not show that we lack obligations to future people, but only that we lack obligations to *particular* future people (Meyer 2010). Our obligations are obligations to *whatever* people happen to exist in the future, and can be derived from the fundamental idea that no person should be deprived by others of the basic resources needed to live and flourish. On this view, if we create a world, through our actions, that leaves future generations without the basic elements they need to survive, then we have failed in our duties of intergenerational justice.

Since much of the literature on intergenerational justice has focused on overcoming a variety of objections to the very possibility of moral duties to future people, robust positive theories of intergenerational justice are few. Nevertheless, several philosophers have developed approaches to intergenerational justice that warrant further consideration and development. Avner de-Shalit (1995) and John O'Neill (1993) suggest that a sense of identity and community with future generations can ground our obligations toward them. Others extend the capabilities approach to justice across time, arguing that we have obligations to support the capabilities of those in the future just as we do for those in the present (see Page 2007). Consequentialist approaches—insofar as they countenance long-term consequences—can also take into account future generations, though there remains the challenge of deciding whether and how to discount the distant future effects of our actions (Partridge 2001, see also Attfield 2003). We might also draw on Kantian ideas of autonomy in developing an ethic for future generations that utilizes a notion of hypothetical consent: Are the choices that we are making now choices that future generations could rationally accept? Finally, the idea of sustainability, discussed in Chapter 5, attempts to bring together obligations to present and future persons by exhorting us to seek policies that "[meet] the needs of the present without compromising the ability of future generations to meet their own needs" (WCED 1987).

Approaches that focus on capabilities and on sustainability draw attention to the fact that in many respects intergenerational justice is a collective and institutional responsibility. Although one can act individually to conserve resources, in order to bequeath to future generations a stock of resources similar to those we enjoy today, we must develop policies and institutions that support the conditions necessary for future generations to thrive. Chapter 5 picks up on this connection between ethics and institutions, exploring their relationship in detail. In Chapter 6, we will explore in more depth questions of climate change and justice, both within and between generations.

For further thought

1. Is the Principle of Commensurate Burdens and Benefits fair? Would another principle be more fair?
2. What institutional changes might improve substantive environmental justice? What changes might improve procedural environmental justice?
3. In your view, what are the strongest arguments for and against the idea that we have obligations to future generations?

Conclusion: Interspecies justice? Anthropocentrism and nonanthropocentrism revisited

In this chapter, we have examined some of the social dimensions of environmental problems. Ecofeminism stresses the way in which systems of domination that exist in the human sphere are tied to and help support similar relationships of domination and oppression in relation to the natural world. Theories of environmental justice examine what constitutes a fair distribution of environmental goods and burdens within and across generations, and explore questions of equity in participation as well as the role of recognition in a just society. The environmental justice movement historically has focused heavily on pollution and the excessive burdens of pollution experienced by women, minorities, and the poor. Because environmental justice community efforts often focus on resisting new environmental burdens and seeking redress for past harms, some environmental philosophers have worried that environmental justice is anthropocentric and thus inconsistent with the fundamental aim of overcoming our human-centered worldviews and extending moral consideration to nonhuman life and the environment as a whole. Tensions between nonanthropocentric environmental ethics and environmental justice have been exacerbated by misanthropic environmental writings, some of which suggest that the planet would be better off without humans in existence at all. Others have argued that ethical obligations to protect the environment may, under some circumstances, trump obligations to feed starving people (Rolston 1996). In some cases, those on each side of the anthropocentrism/nonanthropocentrism debate misunderstand one another, or convey poorly their central ideas. In other cases, we face genuine tradeoffs between benefiting people and benefiting the natural world.

Nevertheless, it is arguably problematic to proceed in environmental moral theorizing without taking into account the deep inequities among people that exist in today's world, as well as the way in which our current policies benefit present persons at the expense of posterity. Conversely, theories of environmental justice that focus exclusively on human beings, and on the distribution of environmental benefits and burdens, risk taking an overly narrow perspective in which the true value of the natural world is reduced to people's immediate needs, preferences, and aversions, leaving out both the significance of our deep dependency on the natural world and the interests of other living things themselves. Although it may never be possible to fully reconcile sustenance of animals, plants, and ecosystems with the interests of individual human beings, efforts to broaden environmental justice to take into account the natural world, along with efforts by nonanthropocentrists to take account of social justice concerns, will be important in navigating the complex relationships between humans and the environment in the decades and centuries to come.

Further reading

Bullard, R. D., Mohai, P., Saha, R., and Wright, B. (2008). "Toxic wastes and race at twenty: Why race still matters after all of these years." *Environmental Law,* 38, 371–411.

Cuomo, C. J. (1998). *Feminism and Ecological Communities*. New York: Routledge.

De-Shalit, A. (1995). *Why Posterity Matters: Environmental Policies and Future Generations*. London: Routledge.

Kheel, M. (2008). *Nature Ethics: An Ecofeminist Perspective*. Lanham, MD: Rowman & Littlefield.

Meyer, L. (2010). "Intergenerational justice," in E. N. Zalta (ed.), *Stanford Encyclopedia of Philosophy*, Spring 2010 edition, http://plato.stanford.edu/archives/spr2010/entries/justice-intergenerational/.

Nussbaum, M. (2000). *Women and Human Development: The Capabilities Approach*. New York: Cambridge University Press.

Plumwood, V. (1996). "Nature, self, and gender: Feminism, environmental philosophy, and the critique of rationalism," in K. J. Warren (ed.), *Ecological Feminist Philosophies*. Bloomington, IN: Indiana University Press, pp. 155–180.
Schlosberg, D. (2007). *Defining Environmental Justice*. New York: Oxford University Press.
Shiva, V. (1989). *Staying Alive: Women, Ecology and Development*. Atlantic Highlands, NJ: Zed Books.
Shrader-Frechette, K. (2002). *Environmental Justice: Creating Equality, Reclaiming Democracy*. New York: Oxford University Press.
Warren, K. (1996). "The power and promise of ecological feminism," in K. J. Warren (ed.), *Ecological Feminist Philosophies*. Bloomington, IN: Indiana University Press, pp. 19–41.
Wenz, P. (2012). "Just garbage: The problem of environmental racism," in L. P. Pojman and P. Pojman (eds), *Environmental Ethics: Readings in Theory and Practice* (6th edn). Boston, MA: Wadsworth, pp. 530–539.

PART TWO

Environmental ethics in practice

5

Ethics, institutions, and the environment

Introduction

Like any other discipline, environmental ethics has its critics, from within the discipline and outside. In thinking through the relationship between theory and practice in environmental ethics, three concerns are particularly important. First, as noted in the Chapter 4, some scholars argue that environmental ethics has focused too much on the nonhuman. At the extreme, critics hold that biocentrism and ecocentrism promote a dangerous misanthropy (Bookchin 1995). Biocentric and ecocentric views need not be misanthropic; however, ecofeminists and social theorists have raised legitimate concerns about the need to examine the interrelationships between human–nature relations and human social relations more broadly. To integrate theory and practice in environmental ethics, it will be crucial to examine and understand the social dimensions of our relations with the natural world, and the way in which "the environment" is bound up with and mediates our social relations.

A second concern focuses on the theoretical orientation of environmental philosophy, with its traditional emphasis on fundamental questions of value. This critique suggests that debates over what possesses—or lacks—intrinsic value have little practical merit. *Environmental pragmatists,* therefore, recommend that we focus not on the foundational values behind environmental decisions, but instead seek consensus at the level of policy and management. According to this view, we can make progress in the realm of environmental practice without settling whether trees possess intrinsic value or not. At a practical level, convergence is possible, despite divergent foundational values. As long as we agree that it is important to protect forest ecosystems, for example, more subtle and fundamental questions about intrinsic value can stay off the table.

There is something important in this critique; however, two replies are worth considering. First, environmental ethics is not *only* about getting things done in the policy arena. It is also about figuring out fundamentally how we should understand, value, and act in the world. Second, sometimes disputes about fundamental values *do* matter. Take, for example, ongoing debates over "ecosystem services." Ecosystem services are "the benefits people obtain from ecosystems" (Millenium Ecosystem Assessment 2003, ch. 2). These include "provisioning services" such as food, "regulating services" such as flood control, "supporting services" such as nutrient cycling, and "cultural and spiritual services," having to do with people's relationship with nature, sense of identity, religious connections, and aesthetic values associated with particular places and the natural world as a whole. There is a strong push toward monetizing and developing markets for ecosystem services as a way to put a price on many aspects of nature that have traditionally fallen outside of economic markets. However, framing our understanding of nature and its value through the lens of "services" is in tension with the idea that we should treat plants, animals, and ecological systems as having value independent of the benefits they provide to human beings. If we choose to value ecosystems only insofar as they sustain human interests and needs, we will develop different policies than we would under a broader conception of value.

Even so, if we are seeking a broadly *practical* environmental ethics, then discussions of abstract value need to link up with the ways in which we live our lives, the policies we set, and the institutions we design. To keep fundamental questions of value in play, a practical environmental ethics will need to consider how institutions embody and reinforce values, and how values inform the design, structure, and functioning of institutions.

Finally, a practical environmental ethics needs to grapple with a third critique, one articulated vividly by Anthony Weston. Weston argues that environmentalists—and his point applies to environmental ethicists as well—have highlighted problems, risks, and dangers, emphasizing the enormity of the challenges we face. He then asks,

> Do we really want to be in the business of overwhelming everyone's last hopes? What happens when we win? (Weston 2012, pp. 2–3)

Environmental ethics has shown where we fall short. Our values are overly narrow. We treat humanity—our own species—as the center of the universe, as all that matters. We have developed oppressive conceptual frameworks that justify both domination of nature and domination of one another. We're cruel to our animal kin. But how about some positive reinforcement? Is there anything that we do well? That we *could* do well? If so, how?

We need aspirations and hope to sustain our motivation. We need an expansive sense of possibility. Or as Weston puts it, we need *vision*. We need to acknowledge what is good in our current values, institutions, policies, and ways of life, and identify openings for change and improvement. We need to dissolve calcified assumptions about the way things are and must be.

Yet vision, too, needs to be grounded. Change often succeeds best when it builds on what came before, when it is aware of the resources—and obstacles—embedded in existing traditions, relationships, policies, institutions, and values. This second half of the book brings into focus some of those resources and obstacles, and in doing so, aims to help us think through the relationship between theory and practice in environmental ethics.

In this chapter, and in those that follow, we look at environmental ethics and its relationship to today's world. This chapter examines the connections between ethics and institutions, showing the reciprocal interplay between them. In Chapter 6, we look at global climate change and the possibilities for a just response to the challenges of a warming planet. Chapter 7 turns to ecological restoration, which may serve as a way of healing the natural world as well as our relationship to it. Finally, in Chapter 8, we take stock of the prospects for a practical and engaged environmental ethics, considering various paths forward for ethical change and moral progress in relation to the natural world.

We will begin, however, by thinking through the relationship between institutions and values, and the constraints and opportunities that emerge from that relationship.

Ethics, institutions, and infrastructure

The word "institution" sounds formal, and even distant from daily life. But the core meaning of the word is quite broad. The *Merriam-Webster's Dictionary* defines institution as a "a significant practice, relationship, or organization in a society or culture." High school graduation ceremonies in the United States, for example, are a cultural institution, as is afternoon tea in England, or *asado* barbeques in Argentina. Institutions can be more elaborate, of course. They include the legal and political systems of a particular society, its educational structure, social welfare programs, financial organizations, and so on.

From birth, we are embedded in institutions. In most parts of the world, tiny babies are soon named, their birth certificates are written, and they enter into citizenship in a particular nation. Our engagement with various institutions often confers various privileges and responsibilities. Institutions are full of norms, laden with values.

Take, for example, educational institutions. Some of the norms and values they promote are intentional and explicit. My children's elementary school emphasizes "teamwork" and "respect" as core values. The school explicitly teaches these values with posters displayed in each classroom emphasizing their importance. Yet, the school imparts other values implicitly, and not necessarily intentionally. Such implicit values—and associated practices—are known as "the hidden curriculum." The lunchroom practices at the school provide an example. All of the lunch trays, tableware, and napkins provided for school lunches are used once, then discarded, creating enormous piles of refuse. Yet, this garbage magically disappears each day, enabling the process to start anew at the following lunch hour. Whether intentional or not, the practice of using throwaway lunch trays reinforces an existing social orientation toward the "convenience" of disposable goods.

To calculate the environmental costs of using disposables instead of reusable plates, glasses, and silverware, we would need to do a full lifecycle analysis of each alternative, looking at the energy required to manufacture each kind of plate, the energy and water used in washing reusable plates, the environmental impact of disposing foam lunch trays, and so on. In general, if plates are reused over their full lifespan and washed in energy- and water-efficient dishwashers, they win out in the environmental calculus.

However, there is more at stake than whether the reusable plates win out in this particular case (though that is not irrelevant). Do we really want to instill in children the values associated with single use and disposal? A recent story on National Public Radio in the United States reported that some Americans apply this ethic not only to food containers, but also to *clothing*. With clothes so cheap, why bother washing them? One person interviewed explained that when he runs out of clean socks, he simply tosses the old ones and heads to the store for a new set. While this example may be extreme, Americans throw out 78 pounds of textiles each year, on average—not including the 1.1 billion pounds of used clothing exported by the United States to other countries (Schor 2010, p. 39).

Returning to the school example, we can see the connections between institutions, values, and *infrastructure*. The most likely explanation for the school's use of disposables is that they are cheaper, or that the school lacks a dishwasher. Let us assume it is the latter, or some combination of the two. To change the institutional practice of using disposable lunch trays will require a change in infrastructure: the addition of a dishwasher, and possibly some plumbing work to connect to the hot water supply. On its own, this does not seem particularly daunting, but other examples demonstrate more prominently the large-scale impacts of infrastructure on behavior and social norms.

Getting to school

Moving beyond the classroom and lunchroom, we can look at patterns of school transportation. In recent decades, travel to school in the United States, where I live, has been transformed by suburban development and greater school choice, the latter of which reflects an effort to reduce racial and socioeconomic disparities in schools. In the East Coast town where I grew up, children were required to attend their neighborhood elementary schools, located within walking distance. Virtually everyone walked or bicycled to school. Today, walking to school in the United States is the exception rather than the rule: less than 13 percent of students in grades K-8 walk to school; in 1969, that number was 48 percent (McDonald et al. 2011). There are various explanations for this trend, but the increase in school choice—an *institutional change* that allows students to attend a school outside of their neighborhood—and development patterns that favor driving rather than walking—an *infrastructure change*—both have been important (Wilson et al. 2010; Napier et al. 2011). Increased parental concern regarding the safety of walking to school also seems to play a significant role (Napier et al. 2011). Moreover, as fewer children walk to school, parents' safety concerns tend to grow, creating a self-reinforcing feedback loop. Walking is no longer the social norm, and fewer walkers mean fewer crossing guards and greater isolation for those who do continue to walk. As parents who are on the fence decide to drive their children, the cycle continues.

This shift in transportation has reduced activity levels in school-aged children and contributes to increasing childhood obesity and negative impacts on health. There are community impacts as well: a study in Ireland found that people in walkable communities know their neighbors better, are more politically and socially engaged, and have greater trust in others as compared to people living in car-dependent suburban neighborhoods (Leyden 2003). As is well known, driving also has major environmental impacts, including local pollution and greenhouse gas emissions. By 2009, 10–14 percent of cars on the road in the United States during morning rush hour were driving children to school (McDonald et al. 2011).

Again, as in the case of the school lunchroom, there are direct and immediate effects of increased driving, as well as an indirect "hidden curriculum" of all this car-based shuttling to school. Here, children learn that even short distances must be traversed by car, setting expectations and patterns for the future. According to one study, among U.S. elementary school children living between a quarter and a half mile from school, only one in four walked (McDonald et al. 2011).

How, one might wonder, do transportation practices, along with the associated institutions and infrastructure that influence them, reflect particular values? Arguably, the shift from walking to driving reflects a priority on speed, efficiency, and parental control. To compensate for the lack of activity built into their children's school days, many parents now shuttle their children to daily after-school sports activities.

Changing these patterns of living requires not only a shift in values, but consideration of the ways institutions and infrastructure can be redesigned to support alternatives. Some communities are pushing back. "New urbanism" seeks to develop walkable communities with centrally located schools and shops. Many towns and cities now mandate that new developments install sidewalks and safe crossings for pedestrians. Others are developing bicycle infrastructure such as bicycle lanes and underpasses below busy streets. At the neighborhood level, parents have organized "walking school buses," where parents take turns walking a group of children to and from school. An elementary school in Boulder, Colorado, makes transportation choices visible by asking students each day how they got to school, recording responses on a chart, and offering small prizes for the classes that rely least on motorized transportation. This practice moves the hidden curriculum associated with transportation choices into the foreground, where children and parents can reflect on their decisions and their impacts. By becoming more aware

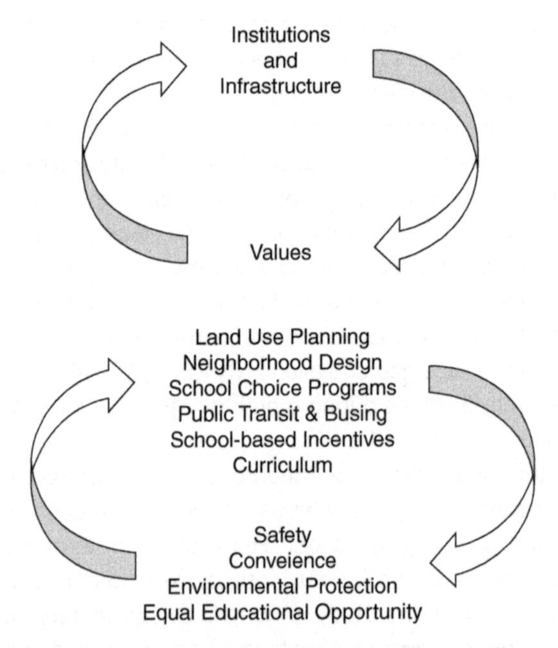

FIGURE 5.1 *Institutions, infrastructure, and values. The top illustration shows the reciprocal relationship between institutions/infrastructure and values. The bottom illustration exemplifies these reciprocal relationships using the example of transportation to school.*

of the values embedded in everyday practices, as well as the constraints and opportunities generated by institutions and infrastructure, people are better positioned to transform institutions in ways that embody values they reflectively endorse.

Of course, the solutions are not always simple, and people often disagree on whether and how to change. School transportation patterns reflect multiple social values. School choice, for example, seeks to enable students to attend the best schools, regardless of where they live. Promoting equal opportunity, diversity, and integration in the educational system is an important value, central to liberal democracy. Can equal opportunity be achieved without compromising other values, such as environmental protection? It certainly seems possible, but finding good solutions will require some creative thought. This is where institutions, infrastructure, values, and vision can all come together with public and political processes of negotiating value disagreements and generating options that communities can support (Figure 5.1).

Eating well: Vegetarianism and beyond

Although many environmental ethicists recognize the connections between ethics and institutions, the field of environmental ethics historically has focused more on value theory than on institutional analysis and critique. There are exceptions, however. Perhaps because it so strongly clashes with values such as reducing animal suffering and conserving and protecting environmental quality, the modern, industrial food system has earned significant attention from philosophers and social critics. Books such as *The Omnivore's Dilemma* by Michael Pollan and Eric Schlosser's *Fast Food Nation*, along with movies such as *Food, Inc.*, have popularized these concerns. These critiques show how our consumer choices implicate us in a complex web of institutions focused on delivering food to our table at low prices, regardless of the effects on animals and the environment.

As noted in Chapter 3, animal rights theorists for decades have highlighted the brutalities of the industrial food system, with philosophers such as Peter Singer and Tom Regan advocating vigorously for vegetarianism. The arguments for reducing or giving up meat are compelling, in light of both ecological and animal welfare concerns. Four decades after the publication of *Animal Liberation*, many farm animals live in similarly abysmal conditions as they did in the 1970s (see Singer and Mason 2006). Chickens live in cramped cages, cows spend the waning months of their short lives standing in their own excrement, shoulder-to-shoulder in huge feedlots, and factory farms confine pregnant female pigs to gestation crates just two feet wide.

In addition, from an ecological perspective, eating lower on the food chain (plants) is more efficient than eating at a higher level (meat). This is because only a small portion of the energy that goes into producing an organism at one level gets transferred up the chain. For example, when a cow grazes on fescue grass, it takes in a portion of the plant—its leaves—and extracts energy from those leaves. But it cannot extract all of that energy, and furthermore, the leaves it eats do not contain all of the energy required to build and maintain the plant in the first place. Similarly, when we eat cows, we obtain some of the energy present in the cow meat, but not all of it. What's more, much of the energy that goes into the production of a cow is not present in the meat. Only a small percentage of the calories consumed by cows contribute to growth; the rest goes to basic sustenance. (For most vertebrates, about 98 percent of energy consumed goes to metabolism, about 2 percent to growth.) Because so much energy is lost from one trophic level (plants) to the next (herbivores), grains, beans, and vegetables take a lot less energy to produce than meat. When one adds to this the fact that animal agriculture is a serious source of nutrient pollution and greenhouse gases, reducing meat consumption seems an obvious choice.

Yet meat consumption throughout the world is on the rise. Meat and animal products now account for half of the daily calories eaten by humans (Blatt 2008, p. 111). In the United States, Americans eat 1 million animals per hour, and the average meat consumption of 218 pounds per person per year by 2003 reflected an increase of 35 percent over the 1950s (Blatt 2008, pp. 112–113).

There seems to be a puzzle here. According to widely shared social values, industrial animal agriculture in the United States is morally problematic. It involves inhumane living conditions for animals and an excess of suffering and death; produces huge quantities of environmentally damaging pollution; and uses vastly more energy and resources than would be needed to produce plant-based proteins for human consumption.

Yet even many people who recognize compelling reasons not to buy industrially produced meat still buy and eat hamburgers, hot dogs, chicken, and steak with barely a second thought, even if they have the resources to do otherwise. Why? Is this pure weakness of the will, a failure to do what one knows is right? Although this may be part of the explanation, it is far from the whole story. For one, giving up meat is not as simple as it might sound: our eating habits are tied to culture, tradition, and identity. To this, one might reply that it is possible to eat meat without eating factory-farmed meat. There are other options. As will be discussed later, not everyone has easy access to alternatives—but assuming one does, why buy meat that is inhumanely produced and environmentally unsustainable?

The answer to these question takes us back to institutions and the ways in which they shape perceptions and choices. The industrial animal agricultural system pushes environmental and animal welfare concerns into the background, foregrounding factors such as convenience, taste, and reasonable price. What we see in the grocery store are neatly packaged cuts of meat, not the processes that went into producing that meat or the pollution and suffering it generated. There is often a tremendous physical and epistemic distance between people and their food, between production and consumption. This in turn facilitates a kind of moral blindness, such that many people do not see or vividly recognize the ways in which their own moral concepts commit them to acting differently.

Although some institutions simply evolve to a size and complexity that make them difficult to understand, in the case of industrial agriculture, much of the blinding is intentional. The animal agriculture industry actively resists witnesses to its practices. In response to activists attempting to make visible to the public the conditions in factory farms, many U.S. states have passed "Ag-Gag" laws that prohibit whistleblowing activities involving the livestock industry. According to these laws, taking video footage of suffering farm animals is now a crime.

The industrial agriculture system is thus ripe for analyses that help us see how social practices and institutions can be morally disabling. In many contemporary agricultural systems, changes in the production context are invisible to the consumer. Eggs look no different whether laid by chickens in battery cages or those that range freely in a local farmyard. Labels such as "natural" and "cage-free" may help, but it is not always clear exactly what they mean. It is easy to choose standard eggs because they are cheaper, the harm involved in their production generally is not visible to us, and it is hard to tell precisely how the (more expensive) alternatives differ. An institutional analysis of food systems would thus examine carefully how their structure inhibits informed consumer choices by limiting transparency and suppressing information.

Although small in scale, community-supported agriculture (CSA) programs offer one response to the transparency problem. In these programs, people have the opportunity to buy "shares" of a local farmer's (or a group of farmers') production each year. Many CSAs encourage their shareholders to visit the farm and observe its workings for themselves, and shareholders often can volunteer on the farm in exchange for a discount on produce. CSAs typically provide a box of vegetables, fruits, and other products such as meat, cheese, eggs, or cut flowers weekly throughout the growing season. Small local farms benefit from CSAs because they provide a committed clientele for farm produce and allow the farmer to vary items in the farm basket throughout the summer, as crops become available. CSA members benefit because they enjoy a steady supply of fresh, local food, typically grown using organic or environmentally sensitive methods. Members can see for themselves whether they support the farm's practices and can engage in direct communication with the farmer about any questions they have.

Farmers' markets represent a similar effort to "localize" the food supply, in part to increase transparency and accountability in the food system, but also to support local economies, reduce food transport distances, and conserve small farms. Urban gardens, where people can rent a plot of land to grow their own food, serve related functions, while building community and bringing people into direct contact with the joys and challenges of planting, weeding, harvesting, and defending one's crops from eager mice, squirrels, and bugs.

Yet not all people can belong to a CSA, buy their food at local farmers' markets, or find the time to grow their own vegetables. The "food movement" in recent years has emphasized organic and local foods, catering strongly to wealthy buyers—and as noted above, food ethics has stressed the environmental and animal welfare impacts of meat consumption. These trends have shaped public dialogue surrounding food, helped develop alternatives to large-scale industrial food systems, and put pressure on the dominant modes of production. However, there are other important issues to consider. As Julian Agyeman (2013, p. 63) notes, "local" is not synonymous with "sustainable":

> The framing of the local food movement in popular discourse has often confused the ends, which are a more sustainable and socially just food system, with the means: the localization of food production and consumption.

Emphasis on the local, for example, draws attention away from fairly traded and sustainably produced foods from other parts of the world, narrowing the opportunities for socially and environmentally responsible farming in developing nations. Similarly, the local foods movement has emphasized environmental concerns over social justice and food access. What's more, not *all* local farms are sustainably run: there is no guarantee that farmers producing food for local consumption will use soil conservation practices or treat their workers better than those producing for people far away (Agyeman 2013). Nevertheless, because of the greater potential for transparency and oversight, local food systems have significant potential, if they explicitly incorporate social justice concerns (Agyeman 2013, pp. 71–72).

Issues of social equity and environmental justice are critical to systemic food reform. While wealthy consumers may be able to afford to purchase high-end organic vegetables and humanely raised meat at their local natural grocer, people living in lower-income neighborhoods, inner cities, or less-developed countries often have little access to fresh vegetables and other healthy foods. Areas that lack sources of reasonably priced, healthy food—often known as "food deserts"—pose an obstacle to healthy eating and the choice of sustainably produced food. In poor neighborhoods, convenience stores and fast food outlets may be the main options, as large groceries, farmers' markets, and natural food stores locate in areas with wealthier clientele. Lack of transportation can further complicate access to good food. A study in New Zealand found that having a car (or not) played a much stronger role in people's food access than the distance to grocery stores (Coveney and Dwyer 2009). Other factors such as price, familiarity, and ease of preparation also influence food choices. Bringing vegetables like kale and collard greens into local markets may not change food selections in any fundamental way if people do not recognize them *as* food or know how to cook them. Food has significant cultural dimensions that are critical to take into account.

Local food policy councils (FPCs) offer one way of broadening the food movement's focus on environmental sustainability to incorporate social justice concerns (Agyeman 2013). FPCs are "collaborative committees that help to coordinate regional food-related activities that strengthen the local economy, the environment, and the community" (Agyeman 2013, p. 73). With diverse representatives from multiple parts of the community, these councils tackle a wide range of issues including food access, transportation, farmland preservation, development of local markets, and public health. FPCs are designed to address local and regional food system issues from a systemic perspective, using inclusive and participatory processes. They gain effectiveness by including representatives from all phases of the food system—from production and processing to distribution and consumption to management of waste—along with concerned citizens (American Planning Association 2011).

FPCs represent an innovation in governance to provide citizens a greater voice in local and regional food systems. In New Mexico, the work of an FPC led to a ban on soft drinks in schools, along with increased availability of healthy fruits and vegetables (Winne 2009). Elsewhere, FPCs have worked to change city ordinances to better accommodate urban agriculture, sought land for community gardening plots, helped improve public transportation to grocery stores and farmers' markets, and developed programs to enable low-income citizens to use federal assistance funds to purchase fresh food at farmers' markets (Agyeman 2013). Like all institutions, FPCs face challenges. Securing funding, developing fully inclusive structures, and even finding mutually agreeable meeting times can pose obstacles to success (Agyeman 2013). However, FPCs in many cities and regions are transforming food institutions and infrastructure in ways that make them more environmentally and socially sustainable. At their best, FPCs draw attention not only to environmental and social justice issues in cities and towns (on the consumer side), but also on the farm. Systemic reform in the use of agricultural pesticides, for example, would address not only chemical residues on foods once they reach the grocery store, but the ways in which pesticides affect farm workers and their families.

In the end, thinking through ways to provide healthy, humane, and sustainable food for everyone involves a complex understanding of the interrelationships among agriculture, economics, the built environment, transportation, and other factors. Not all of these aspects of the food system need to be tackled simultaneously, of course. But the development of institutions that provide a systemic view is a good place to start.

Values, institutions, and infrastructure: From analysis to reimagination

We have now looked at the interplay among institutions, infrastructure, and the environment in multiple contexts. The transportation case illustrated how patterns of suburban development, car-centered infrastructure, and efforts to ensure equal educational opportunity can together contribute to significant changes in transportation to school. Examination of the food system underscored problems with animal welfare, environmental protection, and social justice, and offered some suggestions for institutional reform. By reflecting on institutions, infrastructure, values, and actions, we can see that institutions and infrastructure not only *reflect* our values, but also *shape* them.

Philosopher Albert Borgmann (2006, p. 175) makes this point using a quote from Winston Churchill: "We shape our buildings, and afterwards our buildings shape us." The point generalizes beyond buildings, of course, and it is easy to see in specific cases how this shaping works. When we build roads and highways rather than sidewalks and bike paths, we encourage one another to drive. We develop the habit of driving everywhere we go, and it becomes second nature. Driving is normal; it is no longer something we choose. We forget that there are alternatives. Thus, when we resent traffic, we demand bigger roads. Bigger roads push walking and biking further from view, and make those options more difficult for those who choose them. As Anthony Weston (1992, p. 324) explains:

> The absolute pervasiveness of internal combustion engines . . . is utterly new, confined to the last century and mostly to the last generation . . . Yet, this technology has so thoroughly embedded itself in our lives that even mild proposals to restrict internal combustion engines seem impossibly radical. This suddenly transmuted world, the stuff of science fiction only fifty years ago, now just as suddenly defines the very limits of imagination. When we think of "alternatives," all we can imagine are car pools and buses.

Here is the critical point: institutions and infrastructure not only influence a single decision (like buying hamburgers for dinner on a given day) or even daily habits (like commuting to work by car); they also *channel our imaginations*, narrowing the range of possibilities for change. Stepping back from particular issues—whether food, transportation, suburban sprawl, or toxic pollution—enables us to ask a more fundamental question: How does the contemporary world shape and constrain our imaginations and our sense of possibility? What expectations do we have, and how are those expectations influenced by the places we live and the institutions we occupy? What relationships with nature are made possible by current institutions and infrastructure, from educational systems to financial institutions and workplaces to recreational areas? How do the homes in which we live and the technologies we employ mold our thinking? What opportunities do they open?

In response to these questions, environmental philosophers and others suggest that as we come to inhabit increasingly human-shaped and human-centered worlds, the relationships between humans and nature are being weakened and severed. Writer Richard Louv (2005) suggests that many children now suffer from what he calls "nature deficit disorder." Rather than spend time outside, playing and exploring, kids' lives increasingly revolve around structured activities and electronic media. This, suggests Louv, causes behavioral problems in children, but more fundamentally, it pushes the natural world further and further into the background of human lives. In some cases, the shifts in behavior and experience may be subtle, though significant. College students used to stare out the window when bored in class. Now they pull out their smartphones. People, in general, spend more time staring at electronic devices, and less time looking out the window, and even when we are outdoors, we are arguably less attentive to what we find there. Ironically, some suggest that electronic technologies and social media—though designed to help us connect—further separate people from one another (Turkle 2011).

As the human-dominated world colonizes our imaginations (see Cypher and Higgs 1997), it may become increasingly difficult to appreciate and value the other-than-human, or even to notice it. "Under current conditions," writes philosopher Roger King, "it may be argued that the built environment inhibits our ability to imagine and implement an environmentally responsible world" (King 2000, p. 122). Peter Kahn hypothesizes that people experience "environmental generational amnesia," which means that each generation adapts to an increasingly degraded environment, yet accepts it as the nondegraded baseline (Kahn 2011). Although we may not realize what we're missing, Kahn argues that human lives are diminished by degraded environments and alienation from nonhuman nature, *even if we adapt to degradation and disconnection.* Not only do people suffer direct physical and psychological harms from increasingly polluted air or confinement to windowless indoor cubicles, they may also suffer "harms of unfulfilled flourishing," such as the harm of being unable to move away from human settlements and gain a reflective distance on one's life and interactions with others (Kahn 2011, ch. 12). One might also argue that observing and connecting with nonhuman animals and plants constitutes one dimension of human flourishing: witness the fascination of young children with the diverse forms of life, from insects to squirrels, dolphins to redwoods. Even if we can live without significant contact with the natural world, will we have the intelligence and foresight to protect the critical aspects of it on which we depend, in the absence of robust relationships with nature?

The antidote to a fully anthropocentrized world is not necessarily just more wilderness, where nature is unoccupied and relatively unmanipulated by humans—though such places are important. Anthony Weston (1992, p. 334) suggests that we need physical spaces, within the places humans occupy, that can nurture stronger environmental values:

> Suppose that certain places are set aside as quiet zones, places where automobile engines, lawnmowers, and low-flying airplanes are not allowed, and yet places where people will live. On one level, the aim is modest: simply to make it possible to hear the birds, the winds, and the silence once again. If bright outside lights were also banned, one could see the stars at night and feel the slow pulsations of the light over the seasons.

In certain places, simply increasing our attentiveness to the natural world could help. One of my students, for example, observed 40 species of birds in his backyard and 60 species on the Colorado College campus—and this in the center of a city of over 400,000 people.

The good news, then, is that contexts, institutions, and infrastructure can not only constrain, but also open up new possibilities and new ways of thinking. In designing institutions and infrastructure, we might look to structures that create greater connectedness and a broader sense of possibility in reimagining our relationship with the natural world. We explore some of these possibilities further in Chapter 8. The last section of this chapter considers one promising concept—sustainability—which, despite its many limitations, has the potential to transform our relationships to nature and one another. Sustainability provides an important means by which to reimagine and institutionalize environmental values, in conjunction with social equity and welfare of future generations. As described below, some critical questions are these: Can sustainability be liberating, creative, and visionary? Or will it just reinscribe the same values and practices that have led us into troubled relations with the natural world and one another?

For further thought

1. What kinds of "hidden curricula" have you encountered in your own life and education? What implicit values do these curricula promote? How might institutions and infrastructure reinforce or challenge those values?
2. Explain how food systems in the United States both shape and are shaped by people's values. Can you think of another example of this cycle in everyday life?
3. How might the fact that infrastructure and institutions "channel our imaginations" be problematic for the environment? How might it be positive for the environment?

Sustainability and sustainable development

What is sustainability?

The word "sustainability" suggests continuance (Becker 2012). But continuance of what? Early twentieth century discussions of sustainability focused on sustaining particular resources: sustainable water supplies, sustainable fisheries, or sustainable forestry. In each case, the idea was to determine how best to secure continuity of a particular good for human use. As the century wore on, limitations in this resource-by-resource approach became increasingly clear. For example, fisheries biologists began to recognize that the traditional concept of "maximum sustained yield" presupposed a relatively constant rate of population growth in the target fish species, failing to take sufficient account of population fluctuations or interspecies interactions. Scientists, policymakers, and the fishing industry slowly realized that harvesting any given fish species at the theoretical maximum level in a low population year could be devastating to the long-term sustainability of fish stocks. More generally, the resource-by-resource approach to sustainability began to shift to a more systemic perspective as the connections between development patterns, pollution, deforestation, climate change, and overall ecosystem health grew clearer.

Also during the latter half of the twentieth century, less-developed countries began to industrialize, following in the footsteps of Europe and the United States, and concerns over the long-term fate of natural resources and the environment began to grow. Attention turned to two major global challenges: poverty and environmental degradation. Beginning in the 1970s, the United

Nations (UN) initiated a series of discussions regarding the relationship between the environment and development. Then, in the early 1980s, the UN established the World Commission on Environment and Development (WCED), headed by Gro Harlem Brundtland, the prime minister of Norway.

In 1987, the WCED issued an important report, "Our common future" (also known as the "Brundtland report"), which laid out an important vision and definition of sustainable development:

> Sustainable development is development that meets the needs of the present without compromising the ability of future generations to meet their own needs. (WCED 1987, ch. 2)

The report's vision was at once radical and reformist (Robinson 2004). It was radical in its insistence that environmental issues and poverty be addressed together. The report asserted that environmental challenges could not be overcome without addressing poverty, and that poverty could not be overcome without addressing the environment. By focusing on the deep linkages between poverty and the environment, "Our common future" provided a new lens for thinking about both. However, the Brundtland report's concrete recommendations departed little from the status quo: the report encouraged ongoing economic growth and increasing economic production, and focused on the environment primarily from the perspective of natural resources and human environmental quality, taking an anthropocentric point of view.

The report's definition of sustainable development—meeting the needs of current and future generations—clearly refers to present and future generations of *people*. Thus, the focus of "Our common future" was not on a fundamental shift in values, but on promoting "cleaner" and "more efficient" patterns of development. Humans remained fundamentally at the center, and industrialized Western societies served as the core models for development. For this reason, perhaps, many environmental philosophers have been hesitant to embrace the idea of sustainable development, and the concept of sustainable development has drawn critique from multiple quarters.

Criticisms of sustainability

Although some scholars distinguish *sustainable development* from *sustainability*, in the wake of the Brundtland report the meaning of the two terms became deeply intertwined. The idea of sustainable development laid out in "Our common future" emphasizes sustainability at the global level, foregrounding questions of international equity. However, the basic idea of meeting current needs while taking into account long-term social and environmental consequences remains at the core of sustainability discussions at many scales, from individual institutions (such as corporations or schools) to cities, states, regions, nations, and the globe. Thus, many of the critiques that apply to sustainable development, as articulated in the Brundtland report, also apply to more general conceptions of sustainability. We discuss three critiques below. The first concern is that sustainability and sustainable development are vague—perhaps even vacuous—concepts. A second worry is that sustainability reinforces rather than challenges the status quo. Finally, critics express concerns that sustainability models—such as "the three-legged stool" or the "three pillars" of sustainability—problematically characterize the human relationship to nature.

Sustainability as vague or vacuous

Perhaps one of the strongest recurring critical themes in relation to sustainability and sustainable development is the charge that these terms are vague. For example, what exactly does it mean to "[meet] the needs of the present without compromising the ability of future generations to meet their own needs"? This, of course, will depend on how "needs" are characterized. Should we focus on the most basic of human needs, such as food, water, shelter, and clothing? Or should "needs" be understood as opportunities for education, meaningful work, caring relationships, creative expression, recreation, and so on? One might also wonder about the extent to which sustainability should focus exclusively on *human* needs. Is sustainable development consistent with a radical decline in biodiversity, as long as basic human needs are met? In discussions of sustainable development, there is as yet no consensus about what should be sustained and for how long, or about what should be developed (Kates et al. 2005; Table 5.1).

Sustainability thus can be frustratingly difficult to pinpoint. For this reason, some have argued that the term is vacuous, empty of meaning. In reply, others have argued that we should neither give up on sustainability nor insist on precision in its meaning. On this view, the problem of vagueness is not solved by eliminating it, but by recasting it. Michael Jacobs (1999), for example, argues that attempts to specify a precise meaning for sustainable development miss the point. Sustainable development is a "contestable concept" that plays an important role in political debate about social and economic development. Debates over the meaning of the term are thus "not semantic disputations but . . . substantive political arguments with which the term is concerned" (Jacobs 1999, p. 26). Arguing about sustainability is just what we *should* be doing.

TABLE 5.1 Sustainable development. Debates over sustainable development focus on what to sustain, what to develop, and over what time scale.

What is to be sustained?		What is to be developed?
NATURE Earth Biodiversity Ecosystems	**For how long?** 25 years "Now and in the future" Forever	**PEOPLE** Child survival Life expectancy Education Equity Equal opportunity
LIFE SUPPORT Ecosystem Services Resources Environment		**ECONOMY** Wealth Productive sectors Consumption
COMMUNITY Culture Groups Places	**Linked by** Only Mostly But And Or	**SOCIETY** Institutions Social capital States Regions

Source: U.S. National Research Council, Policy Division, Board on Sustainable Development, *Our Common Journey: A Transition Toward Sustainability*. (Washington, DC: National Academy Press, 1999). Reprinted with permission from the National Academies Press, Copyright 1999, National Academy of Sciences.

We can think about sustainability at two levels (Jacobs 1999). At one level are vague but fairly well accepted characterizations of sustainability, such as that offered in the Brundtland report. At the second level are more precise conceptions of sustainability, which offer specific interpretations of the broader core ideas. At this level, the task is not to *define* sustainability but to work out in practice what sustainability should mean. As an analogy, consider the concept of justice (Vucetich and Nelson 2010). Philosophers and legal scholars do not all agree on a precise definition of justice, but there is a shared understanding of many key issues and questions at stake in the development of theories of justice. Philosophical, legal, and political debates about justice are themselves critical to working out what justice requires, and these debates will continue over time. As Vucetich and Nelson (2010, p. 540) put it:

> Our understandings of justice are varied, indefinite, and evolving. However, by continuing to tend its meaning at all levels of society (i.e., academics, professionals, politicians, and the general public), we have developed viable

> legal systems that evolve with societies' conceptions of justice. [Similarly, achieving] sustainability requires tending its ethical dimension across all levels of society, even though we cannot ever expect to arrive at a final determination of its meaning.

With this in mind, perhaps the vagueness objection to sustainability can be defused by seeing the concept's open-endedness as "constructive ambiguity" that brings together diverse groups to work out in practice a vision for the future (Robinson 2004).

Sustainability as reinforcing, rather than challenging the status quo

Even if one comes to terms with the vagueness and indeterminacy inherent in the concepts of sustainability and sustainable development, one might worry that sustainability does not go far enough in promoting change. There are a variety of ways to express this concern, but the fundamental objection is that sustainability emphasizes minor reforms where radical reorientation is needed. On the ecological side, objectors point to sustainability's anthropocentrism as problematic; while on the social side, critics argue that sustainability too quickly glosses over issues of "power, exploitation, and redistribution" and the need for a fundamental social and political reorientation in order to address them (Robinson 2004, p. 376). Another way of expressing this worry is that discussions of sustainability and sustainable development sometimes seem to suggest that with some minor adjustments, we can continue business as usual. From an environmental perspective, this means continuing to see the natural world merely as a set of resources for our use.

There is no easy answer to this objection. If the response to the vagueness critique is on target, then whether sustainability challenges the status quo or simply reinforces it will need to be worked out in practice, on the ground. To some, this will be unsatisfying, because it leaves a great deal of contingency in the outcomes of sustainability in relation to both environmental protection and social justice. Moreover, because of this flexibility, sustainability often is coopted by vested interests seeking to adopt a mantle of social and environmental responsibility without committing to fundamental, systematic reform.

Sustainability models: A problematic picture of human–nature relations?

A third critique of sustainability builds on the second. This critique focuses specifically on a dominant model of sustainability as a "three-legged stool," or as comprised of "three pillars." According to the three-legged stool model, sustainability rests on the tripod of society, environment, and economy. This model is sometimes described in the related terms of "people, planet, and profit," known as the "triple bottom line." The "3P" model emphasizes the need to broaden the traditional profit-focused perspective so as to incorporate society and the environment as well. Even so, it not clear that the 3P and the three-legged stool formulations of the model are equivalent. Is "economy" the same as "profit"? Is "environment" the same as "planet"? The 3P model focuses more strongly on a business perspective, whereas the three-legged stool or three pillars model considers sustainability more broadly.

Criticism of these models focuses not so much on terminology as on the relations among the three elements of sustainability within them. Although the legs of a stool each play a supporting role in the stability of the stool, they are distinct from one another and each ostensibly plays an equal part. However, critics argue that the environment is not distinct from society or economy, but that society and economy are *embedded within* the environment. As Dawe and Ryan (2003, p. 1459) put it: "[T]he environment is not and cannot be a leg of the sustainable development stool. It is the floor upon which the stool, or any sustainable development model, must stand." In addition to mischaracterizing the relationship between humans and the natural world, by treating each component separately the stool model encourages a mindset of tradeoffs and suggests that sustainability requires balancing each leg of the stool. This can foster oppositional thinking, such as approaches that set jobs against the environment. What is needed, on this view, is a more integrative model that emphasizes the interdependency and relationships among human beings and the natural world. One alternative is an approach that "nests" economy within society, and society within the broader environment (Waas et al. 2011).

Sustainability and future generations

The critiques described above already reveal some of the key fault lines in debates over sustainability, both in theory and in practice. These debates play out more specifically in particular contexts, such as in the debate over sustainability and future generations. This debate introduces a distinction between two forms of sustainability: strong and weak.

We began with the idea that sustainability implies continuance through time, and the Brundtland report emphasizes that sustainability must take account of future generations and their needs. Sustainability calls our attention to questions of fairness and equity, not only among present people, but also in relation to the future. Acknowledging the vague and contested nature of sustainability, Bryan Norton (2005, p. 304) insists that "[s]ustainability, whatever else it means, has to do with our intertemporal moral relations," our moral relations through time. But what does sustainability really demand of us in relation to future people?

One way of thinking about our moral relations through time is in terms of our *bequest* to future generations. This bequest is that which we leave for future people in terms of wealth, resources, institutions, infrastructure, and technology (Norton 2005). From an intuitive perspective, we might think that we should leave the world in no worse a condition than that in which we found it. In relation to the environment, we might interpret this as implying that we should leave the natural world just as it was when we inherited it from *our* forebearers. However, this is an unreasonable demand: the natural world is dynamic, even in the absence of human influence, and trying to maintain a static environment would disrupt natural processes of ecological and evolutionary change.

What seems more plausible, then, is to suggest that we owe to future generations a comparable level of environmental quality or natural resources, or a similar array of healthy ecosystems. But this is where things get tricky. Recall that our bequest to future generations includes not only a certain kind of natural environment, but also built environments, bodies of knowledge, modes of social organization, and a set of technological capacities. With this in mind, consider also that future generations might not want or value the same things as we want and value today. Future people might also have new ways of interacting with the natural world: they might be able to take advantage of natural resources in different ways, or appreciate different aspects of ecological systems than we do now.

Together, these considerations have led some to conclude that we would be mistaken to focus on the natural environment and natural resources per se as the critical elements of our bequest. Instead, we should think in much more general terms about what we owe to the future. From this perspective, defenders of *weak sustainability* argue that we are obligated to provide future people opportunities for an equal level of welfare as we enjoy today; however, *this obligation does not necessarily require any specific sustained level of natural resources or environmental similarity between the present and the future.* While this may seem puzzling at first, consider the possibility that certain goods are substitutable by others. If that is the case, then sustaining a comparable level of welfare over time will not require constant stocks of natural resources. If trees are no longer abundant, we can make paper out of straw. If we run out of clay, we can fabricate bricks from other materials.

There are two ways to object to weak sustainability as just described. First, one might question the premise of substitutability *in practice.* If certain critical resources, for example, have no substitute and their loss cannot be compensated by gains in other areas, then welfare will decline as those resources are depleted. It certainly seems plausible that some resources lack substitutes, but the question turns out to be a complicated one. One might think, for example, that there is no substitute for clean, breathable air. Without it, surely people are worse off. This certainly seems to be the case in many urban areas in contemporary China, where the air at times is so bad that entire cities shut down. A recent study found that current levels of pollution in northern China reduce life expectancies by five and a half years, on average (Chen et al. 2013). However, imagine a new technology that could counteract the health effects of air pollution, such as tiny, implantable respiratory filters that could be placed in the throat to prevent damage to the lungs. With sufficiently sophisticated technology, perhaps we could counter the negative effects of air pollution on welfare. This is the counterargument to the objection that natural resources and environmental quality have no substitute.

However, a second concern about weak sustainability cuts deeper, objecting to substitutability *in principle*. Strong sustainability theorists tend to doubt that the loss of natural resources really can be compensated by greater technology, no matter how good the technology. Or, in the language often used in this context, they doubt that the loss of *natural capital* can be compensated by greater *human capital*. Bryan Norton (2005, ch. 8) suggests that the fundamental divide can be described in terms of "welfare versus stuff." Weak sustainability theorists think that we should leave to future generations the resources they need to maintain nondeclining levels of individual human welfare. Under this view, if natural capital and social capital are easily interchangeable in practice, then our obligation is to pass on an undepleted supply of total capital to future generations. If natural and social capital are not easily interchangeable in practice, then we should pass along a portfolio of both that will provide for nondeclining welfare. The *strong sustainability* position, however, insists that welfare is not all that matters. There are some elements of the natural world—or natural capital—that matter to society in ways that are distinct from considerations of individual welfare. Therefore, the assumption of substitutability is not just potentially problematic *in practice* (e.g., if we can't improve technology fast enough to compensate for declining environmental quality's effects on human welfare); strong sustainability theorists argue that we cannot *in principle* rely on the substitutability of one good for another. We need to save certain kinds of stuff for future generations. As Norton explains:

> The key theoretical divide here is over the question of substitutability of types of capital: weak sustainability theorists assume there are no limits on substitutability among resources and pay attention only to welfare changes,

> whereas strong-sustainability theorists believer there are limits on such substitutions and specify stuff instead of welfare. (Norton 2005, p. 313)

Although a defender of the intrinsic value of nonhuman life would likely endorse strong sustainability on the grounds that we should leave to future generations a world undiminished in value, one need not commit to strong forms of intrinsic value to embrace strong sustainability. Strong sustainability may simply reflect the view that humans value things other than welfare. For example, if people care about the existence of diverse species of animals and plants, and not only because this diversity contributes to individual human welfare, then strong sustainability might support an obligation to protect biodiversity for future generations.

The debates over strong and weak sustainability are complex and not easily resolved, but at the bottom it often boils down to divergent views of what we owe to future generations. Is our obligation to ensure that future generations can be equally wealthy as present people, as weak sustainability would suggest? Or do we owe future generations an undiminished natural world, or at least an undiminished stock of natural capital, as strong sustainability theorists would have it?

A core question that divides weak and strong sustainability views is this: If future generations are likely to be wealthier than us—as optimistic extrapolations from past trends might suggest—then does it really matter that they may live in a world with fewer old-growth forests, less biodiversity, and murkier skies?

Deepening, sustaining and institutionalizing sustainability

Conflicts over sustainability raise critical issues for our thinking about how to conceptualize and institutionalize it. As noted above, weak sustainability is very clearly human-centered, focusing on the continued capacity to provide for human welfare and allowing tradeoffs between natural and human capital to accomplish this goal. Strong sustainability rejects the idea that human and natural capital are fully substitutable. Thus, it requires the protection of at least some forms of natural capital. However, even strong sustainability is compatible with anthropocentrism; it requires only that there are some losses of natural capital that cannot be fully compensated by increases in human capital. Thus, nonanthropocentrists are likely to remain dissatisfied with current, dominant conceptions of sustainability. We see this in certain objections to sustainability, which suggest that what is needed is not a more refined understanding of how to utilize the natural world for human benefit, but a reconceptualization of humans' relations to the natural world (Robinson 2004; Newton and Freyfogle 2005).

As an antidote, or a supplement, to the anthropocentrism of the Brundtland Report, the Earth Charter endorses 16 principles of sustainability that include emphasis on respect for "life in all its diversity," and "care for the community of life" (Earth Charter Initiative). The Earth Charter provides a stronger sense of humans' embeddedness in natural systems (Waas et al. 2011) and a reduced focus on economic growth, while continuing to stress social and economic justice. Although the Earth Charter has not been formally endorsed by the UN, it was produced through an inclusive and participatory process, and is endorsed by over 4,500 governmental and nongovernmental organizations. With its strong emphasis on both social justice and respect for the natural world, the Earth Charter may provide a platform for bringing together two groups of critics of mainstream conceptions of sustainability. Whereas nonanthropocentrists object to an exclusive emphasis on humans and their needs, others worry that sustainability has failed to take sufficient account of social justice (Agyeman 2013). The Earth Charter places a strong emphasis on both the natural world and just societies, and thus may be acceptable to both.

Perhaps, the best way to make a place for a wide range of views is to retain openness in the concept of sustainability and what it requires. This suggestion is consistent with the idea that the constructive ambiguity of the sustainability concept allows for and encourages social and political discussion about what sustainability might mean in particular contexts. Such openness also allows for ongoing evaluation and adjustment of sustainability principles and goals. Sustaining sustainability just may require this kind of process-oriented approach.

For further thought

1. Why is sustainability a potentially powerful concept for environmental ethics? What are its key limitations?

2. How would you respond to the three criticisms of sustainability described above?

3. Which view of sustainability do you find more convincing—weak sustainability or strong sustainability? Why?

4. The Earth Charter is available online at: http://www.earthcharterinaction.org/content/pages/Read-the-Charter.html. After reading the charter, do you think its 16 principles represent an improvement over the Brundtland idea of sustainability? Are there

elements of the Charter that you would change? Elements to which you object?

Conclusion

We began this chapter with a discussion of the relationships among institutions, infrastructure, and ethics. This concluding section links that discussion explicitly to sustainability. Although sustainability at times can seem hopelessly abstract, significant concrete changes can often be accomplished at local scales. What's more, local, regional, national, and international scales are not discrete: they interact in important ways. Take, for example, the American College and University Presidents' Climate Commitment. At the college where I work, students convinced the previous president to sign on to this commitment, which requires institutions of higher education to inventory greenhouse gas emissions and set a date for achieving climate neutrality. The new president inherited this commitment and is now working to ensure that we meet our target climate neutrality date of 2020. Spurred by the Climate Commitment, the college has invested in energy efficiency projects throughout the campus and is seeking to transition from fossil fuel-based energy to renewable sources. In addition, sustainability is now a central pillar in the college strategic plan, and this in turn has generated important conversations about what sustainability means, and how to integrate it into all aspects of campus life. These commitments are changing not only the explicit curriculum at Colorado College but are changing the hidden curriculum as well, by shaping college practices, institutions and infrastructure in ways that convey key values. Students in the college dining hall are expected to compost their food waste, for example, and the campus dining service has committed to serving exclusively vegetarian meals for dinner on "Meatless Monday," in order to reduce energy use and greenhouse gas emissions associated with the production of meat. Students are driving change, too, lobbying for water-wise campus landscaping appropriate for our dry climate. At another local university, a student initiative generated a ban on sales of bottled water on campus: students now bring water bottles and refill them at drinking fountains, reducing energy use and solid waste.

These changes are small, but significant, and can ripple through the community. For example, a college's decision to source local and sustainably produced foods (Agyeman's concerns about the local foods movement notwithstanding) can alter local and regional food markets in ways that favor family farms and greater transparency in agriculture. Fine-scale choices can influence broader scale change, as seen in the nascent fossil-fuel divestment movement on college campuses. Whether or not one views this as the best approach to addressing climate concerns on campus, the divestment movement has generated significant media attention, introducing new considerations into the public discourse about investing in the fossil fuel industry. In this sense, the tactic has been effective in opening up dialogue about energy alternatives, and it has shifted the burden of argument from those who oppose investment in fossil fuels to those who support it. With Stanford University taking a lead role by announcing in spring 2014 its plan to divest from coal, the movement is beginning to gain significant traction. The divestment movement represents one example of an effort to "scale up" social change, moving from local-scale initiatives to broader shifts in institutions and policies. Scaling up is a critical challenge in addressing global climate change, one of our most significant and pressing environmental problems. Chapter 6 addresses climate change and climate ethics, and is followed by a more hopeful chapter on ecological restoration. In the final chapter of the book, we once again return to the issue of social change.

Further reading

Agyeman, J. (2013). *Introducing Just Sustainabilities: Policy, Planning, and Practice*. London: Zed Books.

Borgmann, A. (2006). *Real American Ethics: Taking Responsibility for Our Country*. Chicago: University of Chicago Press.

Earth Charter Initiative. "The Earth charter," http://www.earthcharterinaction.org/content/pages/Read-the-Charter.html.

Jacobs, M. (1999). "Sustainable development as a contested concept," in A. Dobson (ed.), *Fairness and Futurity: Essays on Environmental Sustainability and Social Justice*. New York: Oxford University Press, pp. 21–45.

Kates, R. W., Parris, T. M., and Leiserowitz, A. A. (2005). "What is sustainable development? Goals, indicators, values, and practice." *Environment: Science and Policy for Sustainable Development*, 47(3), 8–21.

King, R. J. H. (2000). "Environmental ethics and the built environment." *Environmental Ethics*, 22, 115–131.

Norton, B. G. (2005). *Sustainability: A Philosophy of Adaptive Ecosystem Management*. Chicago: University of Chicago Press.

Weston, A. (2012). *Mobilizing the Green Imagination: An Exuberant Manifesto*. Gabriola Island, BC, Canada: New Society Publishers.

World Commission on Environment and Development (WCED). (1987). "Our common future." New York: Oxford University Press.

6

Global climate change

Introduction

Global climate change is one of the most serious environmental problems we currently face. It is also one of the most challenging. This is, in part, because the science of climate change is complex and it took several decades to build climate models that could effectively predict many of the major effects of increasing concentrations of greenhouse gases (GHGs). Even now, after the Intergovernmental Panel on Climate Change (IPCC)—an organization that brings together hundreds of scientists from across the world to develop timely and detailed consensus reports on global warming—has issued its fifth report, fine-grained predictions for particular places are difficult to make. Similarly, the exact timeline of various large-scale events, such as melting of the polar ice caps under various emissions scenarios, is difficult to pinpoint.

These uncertainties have led some skeptics to conclude that global climate change is not a matter of serious concern. However, the evidence indicates otherwise. We know with a high degree of confidence, for example, that the earth's atmosphere exhibits a significant warming pattern that corresponds with increasing concentrations of carbon dioxide and other GHGs in the atmosphere. Scientists are observing accelerated rates of glacial melting and rising sea levels, consistent with the predictions of global circulation models designed to simulate the effects of increasing GHG concentrations in the atmosphere. What's more, record-setting temperatures throughout the world in recent years, extended hurricane seasons, and an increasing number of large-magnitude storms corroborate climate model forecasts of storm intensification and weather extremes. There is now an overwhelming scientific consensus that climate change is happening, that it is caused by human activities, and that its impacts are becoming increasingly significant.

Yet, it is difficult to predict precisely what the impacts of climate change will be in any particular place 10, 50, or 100 years from now. The climate is a complex system, with both negative and positive feedback loops. We do not fully understand how these feedbacks work, or how they will interact with one another. We also do not know how to project GHG emissions into the future. The main driver of climate change—anthropogenic emissions—depends on politics, law, economics, culture, and other dimensions of human behavior and social organization. Thus, although we can be highly confident that anthropogenic climate change is occurring, and that it is causing changes in global mean surface temperature, glacial retreat, melting of polar ice, sea level rise, and changes in patterns of precipitation, seasonal snowmelt, river flows, and storms (IPCC 2007b, 2007c; Karl et al. 2008), the precise magnitude of these effects will depend in complex ways on the choices we make. As discussed below, this and other features of global climate change make it a particularly thorny problem from an ethical perspective. Before turning to that, however, it will be helpful to have a brief background on the science of climate change, and the diverse lenses through which people view the problem.

Climate basics

Earth's climate has changed radically throughout the planet's history. Seas once covered areas that are arid semi-deserts today, and places previously buried under thick layers of ice now host temperate forests. Contemporary climate change is distinctive, however, for three reasons. First, it is *human caused.* Second, it is occurring at an *extremely rapid rate*, and third, its *magnitude* is significant. We currently face climate change over the course of the next century "that is comparable in magnitude to that of the largest global changes in the past 65 million years but is orders of magnitude more rapid" (Diffenbaugh and Field 2013). For plants, animals, humans, and ecological systems as a whole, this poses tremendous challenges. Recent research suggests that in order to adapt to predicted climate changes between now and the year 2100, mammals, birds, and amphibians would need to evolve more than 10,000 times faster than the historic norm (Quintero and Wiens 2013). Species can employ another strategy for coping with a warming climate, shifting their ranges toward the poles or to a higher elevations, however such shifts will be difficult in many areas due to habitat fragmentation by human development (Quintero and Wiens 2013; Diffenbaugh and Field 2013).

These sobering predictions prompt us to ask why the climate is changing so quickly. The simple answer is that humans have been emitting GHGs at rapidly increasing rates since the time of the Industrial Revolution, when we began to burn large quantities of fossil fuels such as oil, coal, and natural gas. *Greenhouse gases* are heat-trapping gases in the atmosphere. They are permeable to short wavelengths of energy, such as light, but relatively impermeable to the longer wavelengths of heat energy. Thus, GHGs enable light energy from the sun to enter the earth's atmosphere, where it warms the surface of the planet. However, when the earth reradiates this warmth, GHGs prevent that a portion of that heat energy from escaping back into space.

GHGs such as carbon dioxide and water occur naturally and play an important role in regulating the earth's temperature. Without them, the earth would be a cold and lifeless planet. However, human emissions of GHGs such as carbon dioxide are changing the earth's energy balance, driving the planet toward a warmer state.

At Mauna Loa Observatory in Hawaii, scientists have been tracking carbon dioxide concentrations in the atmosphere since the 1950s, and a graph of emissions over time shows significant increases from 315 parts per million (ppm) during the mid-twentieth century to more than 400 ppm in the spring of 2013, *the highest level in 3 million years* (Gillis 2013). This increase exceeds the pre-Industrial Revolution baseline of 280 ppm (IPCC 2007b) by more than 40 percent, and has been generated primarily by the combustion of fossil fuels for heat, electricity, and transportation. It is important to note that although carbon dioxide is perhaps the best-known GHG (witness organizations such as 350.org, whose focus is to stabilize carbon dioxide levels at 350 ppm, or proposals for "carbon taxes" or "tradable carbon credits"), other gases play an important role in climate change. Methane, for example, is produced not only by natural decomposition, but by livestock, natural gas leaks, and landfills. Although human activities produce significantly less methane than carbon dioxide, methane is a more potent GHG: per pound, methane has a warming effect more than 20 times that of CO_2.

The most obvious and direct effect of increased GHG concentrations is overall warming, leading to rising global mean surface temperatures. However, anthropogenic global climate change is causing various other effects. Some are fairly obvious, like melting glaciers, melting sea ice, and rising sea levels, while others are more subtle and indirect. For example, loss of sea ice changes the *albedo*, or reflectivity, of the ocean. Since darker surfaces (open ocean) absorb more solar energy than lighter ones (sea ice), reductions in sea ice can generate feedback loops that accelerate ocean warming and spur further sea ice melting. Climate change is also associated with altered patterns of wind and cloud formation, changes in rainfall, and shifts in the frequency and intensity of forest fires. In marine systems, elevated levels of atmospheric CO_2 are changing ocean chemistry. When carbon dioxide dissolves in the ocean, it forms carbonic acid, decreases pH, and reduces the availability of calcium carbonate, which many marine organisms need to make shells or skeletons (NOAAa).

The effects of climate change on weather, climate, and ecological systems also clearly influence human beings and their livelihoods. As sea levels rise, coastal lands will be inundated and low-lying islands may be entirely submerged. Already, declining sea ice in Alaska has exposed coastlines to stronger waves and rapid erosion, such that many native villages are facing relocation. The native people of Shishmaref, Alaska, for example, may need to evacuate their village on the Chukchi sea, which they have inhabited for the past 400 years (NOAAb). The changing patterns of rainfall due to climate change also may make agriculture more difficult in many areas, and higher intensity storms, increased flooding, and more frequent heat waves put people all over the world at risk. Although locally positive effects may occur—agriculture in some areas may benefit from a longer growing season, for example—in general, climate change is increasing stress on natural and social systems. Throughout Africa, for example, temperatures are increasing and precipitation patterns are changing; in the coming century, these shifts will likely exacerbate "existing health vulnerabilities . . . including insufficient access to safe water and improved sanitation, food insecurity, and limited access to health care and education" (IPCC 2014, ch. 22). These changes, in turn, may lead to changes in human population and migration patterns, greater competition for scarce resources, and increased armed conflict.

Given these concerns, it seems obvious that climate change is a serious problem demanding concerted action. Why, then, does climate change remain so controversial? Why is there so much disagreement about how to address this growing challenge?

Why do we disagree about climate change?

In his book, *Why we disagree about climate change*, Mike Hulme (2009) offers several explanations for the vigorous and often polarized debates about climate change. First, climate change is framed in many different ways, through science, economics, religion, ethics, and risk. Despite all the scientific discussion of "parts per million," global circulation models, and atmospheric chemistry, climate is not only a physical phenomenon, but also a social and cultural one. The way we perceive and interact with climate depends very much on the social, cultural, and ethical meanings that climate holds for us. As Hulme (2009, p. 2) explains:

> There may be "good" or "benign" climates and "bad" or "dangerous" climates, but only in the sense that climates acquire such moral categories through human judgements—judgements that suit our convenience or our capabilities.

Although we often attempt to separate the science of climate change from its political, economic, and ethical dimensions, climate science is bound with society in many important ways. As discussed in Chapter 1, science itself is inherently value-laden. The social, political, economic, and ethical dimensions of climate change provide a context in which climate science takes place, and the very existence and importance of the IPCC and its scientific assessments clearly reflect this context.

Despite this, we often unreasonably expect science to settle climate debates. Thus, "one of the reasons we disagree about climate change is because science is not doing the job we expect or want it to do" (Hulme 2009, p. 74). Where we expect simplicity, science delivers complexity, and where we want science to tell us what to do, we find that it cannot, at least not by itself. Deciding what to do requires deliberation, negotiation, and weighing of values. *Even if* we could accurately predict the precise outcomes of climate change for each part of the world under different policy scenarios, we would likely still disagree about what to do.

Deciding how to address global climate change is particularly difficult because standard decision-making tools such as cost–benefit analysis do not work well in the climate case. Common concerns about cost–benefit analysis—that it overlooks costs and benefits that lack a market price, and that it attempts to compare incommensurable goods on a common scale—pose problems for cost–benefit analyses of climate policies. However, additional challenges arise due to the uncertainty of climate predictions under different emissions scenarios. Without a clear understanding of costs, benefits, and risks associated with different decision pathways, cost–benefit analysis breaks down. This is because cost–benefit analysis relies on quantification of risks based on their *magnitude* and *probability*, and costs cannot be accurately assessed without a thorough understanding of the relevant risks. In the case of climate change, we do not know all the risks, and we cannot accurately quantify all those risks that are known. On this basis, some have argued that cost–benefit analysis is a poor tool for climate planning and decision-making, and that an alternative approach, such as one based on the precautionary principle, would help identify a path that could avert the most severe outcomes. (As discussed in Chapter 2, the precautionary principle suggests that we take measures to protect human and environmental health when our actions place them at risk, even if the precise nature and magnitude of the risks are unknown.)

Cost–benefit analysis of climate change also faces difficulties due to its long time frames. This challenge came to light in a highly publicized debate between economists Nicholas Stern (2007) and William Nordhaus (2007, 2008). When Stern and Nordhaus conducted independent analyses of climate change costs and benefits, they came to radically different conclusions. Nicholas Stern, author of a lengthy report on climate change commissioned by the British government, recommends immediate action on climate change equivalent to an investment of 1 percent of the global gross domestic product (GDP) in order to stabilize CO_2 concentrations at 500–550 ppm (Stern 2007). In contrast, Nordhaus (2007) argues that we would be better off making much smaller investments in mitigation in the near term, and then increasing our efforts by mid-century.

Why do Stern and Nordhaus disagree about climate change? The answer is that they have different views about how to value future costs and benefits—and this in turn reflects fundamental *ethical* differences in their approaches. More specifically, the "Stern–Nordhaus debate" reflects a basic disagreement about the appropriate *discount rate* to use when assessing climate impacts. *Discounting* is a method that places greater weight on near-term costs than on long-term costs, and the *social discount rate* determines the degree to which present costs and benefits are emphasized over future ones. At high discount rates, costs and benefits in the relatively distant future (100 years from now, say) have very little weight. In contrast, without discounting, a human life saved today gets no greater weight than a human life saved 500 years from now. The practice of discounting raises a key question: Are we ethically justified in placing less value on the lives of some people than others, just because they live at a different time (in the future rather than the present)? If so, how much more value should we assign to present people as compared to those who will come after us?

It is tempting to appeal to empirical data to settle the question about whether and how to discount the future. If we look at the observable market behavior of people and the decisions they make, we can infer the degree to which people discount the future in their economic decisions. However, as Stern and Taylor (2007, p. 204) explain, "There is no real economic market that reveals our ethical decisions on how we should act together in the very long term." In other words, market behavior reveals how people make decisions to promote their individual interests in the *short term*, but says relatively little about how people value the long-term future.

In addition, Stern and Taylor make a fairly basic ethical point: what people *actually* do does not necessarily reflect what they *should* do. The fact that cheating is highly prevalent, for example, does not settle the question of whether cheating is morally acceptable. Thus, the *descriptive* discount rate (the rate that reflects actual market behavior) cannot settle the ethical question of what discount rate we *should* use, or how much ethical importance we should grant to future generations. This point is underscored by the fact that descriptive discount rates are based primarily on people's decisions involving the relatively near-term future. So, even if we were justified in using descriptive discount rates to estimate near-term costs and benefits, further argument is needed to justify the application of these rates to time scales of a century or longer.

In the end, we disagree about climate change because climate has social, ethical, cultural, economic, and scientific dimensions: it is not a problem that can be objectively described and solved within the confines of science, and when we attempt to quantify costs and benefits, we find ourselves grappling with complex ethical issues. As we shall see in the following section, these issues are perhaps particularly difficult for us to manage because they stretch the boundaries of our common moral concepts and understandings.

For further thought

1. Describe some of the major effects of global climate change. How might these effects influence your life? How might they influence the life of someone living in a very different part of the world?
2. How does contemporary climate change differ from prior periods of climate change in the Earth's history?
3. Why is cost-benefit analysis difficult in the case of climate change?
4. Is discounting the value of future lives ethically acceptable? Why or why not?

The moral challenges of climate change

At an intuitive level, it seems obvious that anthropogenic climate change is a moral issue. It raises questions of *harm* as well as questions about the *fair distribution of burdens* in preventing and addressing harm. Let us start with harm. Climate change is clearly harming humans, other living things, and ecological systems. All of these harms are of concern. In addition, humans are causally responsible for the harms associated with our GHG emissions. Although causal responsibility does not always imply moral responsibility, in the climate case, we are aware that our activities are causing harm and there are steps we can take to reduce it. Thus, although we disagree over how to apportion moral responsibility for climate change, it seems clear that we *are* morally responsible.

This brings us to the second key moral issue of climate change: *the fair distribution of burdens.* International climate negotiators have been grappling with the issue of fairness for decades. How should we distribute the burdens of reducing GHG emissions? Which countries should make the biggest emissions cuts, and why? Should less-developed nations be permitted to develop along the same fossil fuel-intensive paths as their industrialized counterparts? Who is morally responsible for climate change, and what does this responsibility require in response?

Although it is widely agreed that climate change raises issues of both harm and fairness, there is controversy about the nature and scope of our moral responsibilities. Philosopher Dale Jamieson (2010) argues that the ethics of climate change is difficult, in part, because *climate change is a nonparadigmatic moral problem*. In other words, climate change does not fit our standard ethical models and thus raises special difficulties in identifying moral wrongs and allocating responsibility for them.

Climate change as a nonparadigmatic moral problem

Let us turn to some interesting and illustrative cases to show how climate change stretches common moral understandings (Jamieson 2010). Imagine a girl named Jill, whose bicycle is stolen by a boy named Jack. What are the features of this case? There is a clear individual harmed (Jill) and a clear individual responsible for that harm (Jack). The nature of the harm is also clear—Jill has been deprived of her bicycle—and Jack obviously *intended* to steal it. There is no significant gap in time or space between Jack's action (stealing the bike) and the harm Jill suffers (deprivation of her bike). This is a *paradigmatic moral wrong*, and Jack is obviously morally responsible for what he has done. But let us look at some further examples, as described by Jamieson:

> In Example 2, Jack is part of an unacquainted group of strangers, each of which, acting independently, takes one part of Jill's bike, resulting in the bike's disappearance.
>
> In Example 3, Jack takes one part from each of a large number of bikes, one of which belongs to Jill.

We follow these examples all the way to a sixth:

> [A]cting independently, Jack and a large number of unacquainted people set in motion a chain of events that causes a large number of future people who will live in another part of the world from ever having bikes. (Jamieson 2010, p. 436)

Jamieson argues that this last example most closely mirrors the moral relationships involved in climate change. Here, the harm Jack causes is indirect, and insofar as his actions cause harm, they do so only in concert with the actions of others whom he does not even know. There is neither a collective plan to deprive future people of bikes, nor is it Jack's intention to do so. What's more, the harm caused is distant in time and space, involving unknown individuals who do not yet exist.

This last example does resonate in relation to climate change. When I crank up my thermostat in winter to keep warm, I have no intention of harming anyone. It is only in virtue of the fact that many other people use fossil fuels to heat their homes (and have done so for more than a century) that my fossil fuel use contributes to the problem of anthropogenic global climate change. What's more, when I turn up the thermostat today, the climate effects are not immediately felt. There is a time lag between cause and effect. The climate system is relatively slow moving (for now) and the impacts of one individual's GHG emissions are not immediately noticeable or traceable. This makes it impossible to assign causal responsibility to particular individuals for particular climate-related harms, and the lack of intent seems to complicate moral responsibility as well.

In Jamieson's view, these characteristics stretch our understanding of moral responsibility, particularly at the individual level. Take the example of driving, just for fun, on a sunny Sunday afternoon (Sinnott-Armstrong 2005). If I do not *intend* to harm anyone by driving my car, and my driving is harmful only because several other people are doing something similar, and many of the people who will be harmed do not yet exist, then how can I really be doing something *wrong* by driving? Is the harm I cause by driving something for which I am morally responsible? Because global climate change is not a paradigmatic moral problem, we are uncertain about how to answer these questions and thus lack a sense of moral urgency surrounding it.

Climate and moral psychology

By drawing the connection between our conceptual understanding of climate change as a moral problem and our motivation (or lack of motivation) to act, we can see that global warming is an ethical issue with important psychological dimensions. It raises not only questions of right and wrong action, fairness in distribution, or assignment of moral responsibility, but also questions of *moral psychology*. Although Socrates famously held that moral knowledge is sufficient for moral action (if you know what's right to do, you'll do it), it is clearly more complicated than that. In the case of climate change, the very features that make the problem morally challenging to grasp—its global scope, it dispersed causes and effects, the time lags between emissions and their climatic consequences, and implications for future generations—make us prone to "moral corruption" (Gardiner 2006). Moral corruption involves psychological strategies—rationalization, selective attention, and self-deception—that lead us to avoid facing fully the moral challenges climate change poses. As Stephen Gardiner explains,

> [T]he complexity of climate change] may turn out to be *perfectly convenient* for us, the current generation, and indeed for each successor generation as it comes to occupy our position. For one thing, it provides each generation with the cover under which it can seem to be taking the issue seriously—by negotiating weak and largely substanceless global accords, for example, and then heralding them as great achievements—when really it is simply exploiting its temporal position. (Gardiner 2006, p. 409)

In other words, it is more appealing to deal with climate change tomorrow rather than today, and it is tempting not to deal with it at all. At one level, we recognize that climate change is a serious problem—yet at another level, we want to deny it (see Norgaard 2011). To continue along our current trajectory is vastly easier than working to transform our modes of transportation and energy use, infrastructure, and patterns of consumption.

Moreover, it is not just greed that drives complacency and moral corruption in the climate case (though this is not to say that greed is never involved). Our current ways of life support many genuine goods, particularly for those of us in wealthy, industrialized nations. In the United States, for example, many people have access to a wide variety of foods year round, live in comfortable homes, enjoy access to a vast array of ideas and information through the internet, and so on. Cars enable a weekend trip to the city or the mountains, and shuttle children to music lessons and sports events; airplanes make possible visits to relatives on the other side of the world.

In the case of global climate change, the activities that cause harm are not *intended* to do so, and they may even be necessary for us to live as we have come to expect. Given current institutions and infrastructure, we need to emit GHGs in order to live. In this sense, we cannot take care of our own needs without contributing to global climate change and to climate-related harms. Our situation, as individuals, may thus be one in which it is impossible to promote all of our core values simultaneously. The necessity of emitting GHGs may excuse us to some degree, but it also suggests that we should do everything we can to transform our situation. Ideally, it should be possible to sustain ourselves without generating significant harm to people and ecosystems, now and in the future.

Climate and collective action

In discussing climate change as a conceptual, ethical, and psychological challenge, we have begun to consider questions of individual moral responsibility. But what, exactly, *are* the moral obligations of individuals in relation to climate change? Environmental philosophers tend to agree that individuals have obligations in this realm, but disagree over the specific nature of these obligations. Intuitively, a reasonable first effort at defining individual obligations might focus on equal per capita allocations of sustainable global GHG emissions. An obligation along these lines might be grounded in the idea that each person should do their fair share, and that each of us should act in ways that others could rationally accept.

Complications arise, however, for an equal per capita shares approach. For one, infrastructure and institutions constrain the emissions needed in order to live a minimally decent life. Thus, an individual's GHG emissions are not entirely under his or her own control (Baatz 2014). I can choose whether to turn on the lights in my office, but not how the electricity that powers those lights is produced. If generated by a coal-fired power plant, each kilowatt of electricity creates a significantly larger carbon footprint than if generated by water, solar, or wind. An individual's GHG emissions are thus a function not only of electricity used, miles traveled, or products consumed, but of the infrastructure and technology that supports that electricity production, transportation, and consumption. Depending on where you live, it may be impossible to reduce your GHG emissions to a sustainable per capita level without truly radical adjustments in your lifestyle. In order to meet the limits set by an equal per capita approach, "the average European citizen has to reduce her emissions by more than 80% minimum (sixfold reduction) and for the average US-American citizen it would be 90% minimum (tenfold reduction)" (Baatz 2014). In light of these complications, Christian Baatz (2014) argues that an equal per capita emissions approach is unjustified. On his view, each of us has a duty to curtail our personal GHG emissions, but this duty cannot be precisely defined. Although we should avoid excessive emissions, what qualifies as "excessive" depends on each individual's particular situation and capabilities (Baatz 2014). The worry, of course, with such an approach is that it invites the very psychological failings discussed earlier: if our duties are open-ended and only vaguely defined, it may be all too easy for us to sidestep them, or rationalize away any robust obligation to change.

There are other ways to define an individual's obligations to reduce emissions, but let us turn now to the more radical claim that *there are no such individual moral obligations*, at least in the current context (Johnson 2003; Sinnott-Armstrong 2005). In developing this idea, philosopher Baylor Johnson calls our attention to an important feature of anthropogenic climate change: in many ways, it resembles a classic collective action problem called *the tragedy of the commons* (Hardin 1968). Johnson believes that this carries significant implications for individual responsibility.

Collective action problems arise when there is a tension between individual interests and the promotion of some collective good. The classic case is that of a grazing commons, where individuals put their sheep out to pasture. If the commons is unregulated, then each farmer has an incentive to add more sheep to the pasture, in order to maximize their own production. Yet if every farmer continues to add sheep without limit, the pasture will be overgrazed and the resource will collapse. This is the tragedy of the commons.

Standard collective action problems can be difficult to solve. But climate change is not a standard collective action problem, and various features make it even more difficult. First, climate change is *global in scale*, thus the number of actors involved is many orders of magnitude larger than in a local case (see Gardiner 2006; Ostrom 2010). Second, there is a *temporal lag* between GHG emissions and their effects on global climate. Third, we lack established institutional structures for managing collective action problems at global and intergenerational scales, and (remember Jack and Jill) we lack *established conceptual frameworks* for handling the moral issues associated with dispersed causation, dispersed effects in time and space, and our obligations to future generations. For these and other reasons, philosopher Stephen Gardiner (2006) calls climate change "the perfect moral storm."

In response to this situation, one might think that each person ought to do his or her fair share, "[restricting] his or her use to a level that would be sustainable if all other users reduced their use in a similar way" (Johnson 2003, p. 272). We have already seen that this formulation does not account for the diverse social, institutional, and technological contexts in which individuals are embedded. However, Johnson believes that there is a more fundamental problem: he argues that *the very idea of individual emissions reductions is misguided* in the current context. In an unregulated commons, one individual's restraint can be exploited by others. If I give up my car and stop buying gasoline, I do nothing to slow global climate change, according to Johnson. My fossil fuel diet simply enables another's fossil fuel binge, because I have left more resources for others to exploit. If this is right, then fossil fuel asceticism at the individual level makes no sense: my restraint has no positive effect.

From these arguments, Johnson concludes that individuals have obligations to reduce emissions *only when there exists a collective agreement to do so.* If we have climate-related obligations, then, they are obligations to create a collective agreement.

There is something paradoxical about this position, however. For one, there is a tension between the obligation to work for collective restraint and the lack of obligation to work toward individual restraint. One way of describing this tension is in terms of *integrity.* Integrity is a virtue that calls for coherence among one's various projects and commitments, as well as the integration, or internalization, of one's values and commitments into one's identity (Audi and Murphy 2006, Hourdequin 2010). A person of integrity seeks coherence among her values and consistency between her values and actions. Although such consistency may help promote certain goods—she is more likely to be respected as an environmental activist working for sustainable transportation if she walks, bikes, or takes public transit to work—integrity reflects a deeper commitment to one's values independent of their consequences. Integrity might lead a person committed to justice to protest a racist government policy, even if he knows his protest is unlikely to succeed. An individual of integrity "wants or chooses to be a certain kind of person, or to live a certain way of life, [such] that her life loses its meaning or point if she fails to actively fight against what she believes to be a demeaning and unjust institution" (Halfon 1989, p. 146).

Integrity arguments may not convince committed consequentialists, however, unless integrity itself promotes good outcomes. Consequentialists, therefore, may find a second reply to Johnson's position more compelling. This reply points out that social change is often catalyzed by efforts at multiple scales. If I stop using my car and start taking the train to work, for example, my neighbor might be intrigued and ask about the train, or even try it out, especially if I speak highly of the experience. Individual actions—especially ones that are visible—can catalyze a cascade of further changes (cf. Knights 2012, ch. 5).

Individual choices also change the relationships among individuals, institutions, and policies. If I am committed to reducing my own ecological footprint, I will be more attentive to policies and institutions that facilitate or inhibit my doing so. My efforts to save water, for example, might lead me to lobby for changed laws to permit the recycling of gray water (water from dishwashing or laundry) for landscape use, or to allow the installation of composting toilets. Another example is this: if I purchase photovoltaic solar panels for the roof of my house, I am likely to have a greater interest in the energy company's policies surrounding electricity metering. It now matters to me not just in principle whether the utility company pays retail rates for electricity my home feeds back into the electric grid; it matters to me personally. This may give me the extra nudge needed to argue for "net metering" policies that allow customers to receive full credit (at retail prices) for electricity returned to the grid.

Such feedback relationships can work at other scales as well. For this reason, Elinor Ostrom (2010), winner of the Nobel Prize in economics, has advocated a *polycentric approach to climate change*. Rather than focus our efforts and our hopes on a single, international climate agreement, Ostrom recommends that we work at multiple levels and take advantage of synergies between them. Individual emissions reductions help open the path for initiatives at larger scales, and contribute to a culture that takes climate change seriously. This, in turn, may help to overcome the socially reinforced denial that impedes our progress.

There are thus convincing arguments in favor of an obligation to reduce one's personal GHG emissions, but we can take two important points from Johnson's position. One is that reducing your own emissions is not enough: to catalyze social change, individuals also need to work to reform institutions and policies. Obligations of integrity go two ways: they imply that individual emissions reductions should accompany social and political action, and that social and political action should accompany individual reductions. What's more, if we return to the relational approach to environmental ethics discussed in Chapter 3, we can see that the individual versus political dichotomy begins to dissolve. As relational beings, we can engage in personal reductions that are also social and political actions (Hourdequin 2010).

At the same time, climate change poses challenges that call for coordinated institutional change. Moving from the individual to the institutional level, we again face an important question: What would constitute a fair division of responsibility in mitigating and adapting to global warming? More specifically, we need to consider how nations should divide responsibility for emissions reductions and the costs of adapting to a changing climate.

For further thought

1. In what respects is climate change a non-paradigmatic moral problem? In what respects is it a paradigmatic moral problem?
2. In your view, does every person have a moral obligation to reduce his or her personal greenhouse gas emissions? Why or why not?
3. Why is global climate change a collective action problem? Why does this matter, from an ethical perspective?

Justice and climate change

Although scientists began to develop a basic understanding of GHGs in the nineteenth century—and even at that time suggested the possibility of human-caused modification of the global climate—it was not until the end of the twentieth century that understanding coalesced sufficiently to gain political attention. By 1988, the IPCC was up and running, and in 1990 they issued their first report, affirming that the international community should take action to address the threats posed by global warming.

This, in turn, helped catalyze development of the United Nations Framework Convention on Climate Change (UNFCCC), established in 1992. This treaty acknowledged the problem of anthropogenic global climate change and the need for cooperative international action to address the threat. The UNFCCC established that developed nations should take the lead in reducing GHG emissions and set a target for reductions by these countries. The initial goal was to reduce emissions to 1990 levels by the year 2000. Although these targets were not fully met, 195 countries now have ratified the treaty, and the parties to the convention have met regularly during the last two decades to negotiate further climate strategies and agreements. In the meantime, scientific reports of the IPCC have expressed with increasing certainty the imminence of anthropogenic climate change and the severity of its effects. Despite greater scientific understanding and regular meetings of the parties to the framework treaty, negotiations have repeatedly stalled, due in significant part to disagreements over what a fair division of responsibility might be.

In deciding how to divide responsibilities for climate change, initial negotiations focused primarily on *climate mitigation*, or reducing net emissions of GHGs. However, as years passed with little progress on mitigation, it became increasingly clear that climate change would not be stopped in its tracks. (Even if we stopped emitting GHGs today, we would be committed to significant climate warming over the next century, due to the long lifetimes of GHGs in the atmosphere and the lag between increased GHG concentrations and their climate effects.) Thus, negotiations shifted to encompass not only mitigation, but also *adaptation*: strategies to cope with the effects of climate change. Adaptation may involve a range of approaches, such as building sea walls, increasing storm preparedness, developing new crops adapted to warmer, wetter, or drier climates, and even relocating towns, cities, and possibly entire populations of small, low-lying nations.

In addition to adaptation, a third response to climate change has gained greater attention in the last few years. Scientists, engineers, think tanks, politicians, and others are now exploring the possibility of *geoengineering*, the intentional manipulation of the earth's climate to counteract the effects of dangerous climate change. Those who favor geoengineering often argue that neither mitigation nor adaptation will be sufficient to prevent very serious climate impacts, and that we need new tools to avert a crisis. Geoengineering is a relative newcomer in international climate discussions, and it raises ethical concerns discussed separately below.

With the main responses to climate change on the table, we can now turn back to questions of justice. In determining a fair distribution of responsibility for climate mitigation and adaptation, wealthy, industrialized nations such as the United States have found themselves frequently at odds with poorer and less-industrialized nations such as India and China. From the beginning, the UNFCCC distinguished two groups of countries—Annex I (developed) and Annex II (less developed/developing)—and established differentiated responsibilities for these two groups. Philosophers working on climate issues generally have supported differentiated responsibilities, for three key reasons:

1 the industrialized nations have historically contributed the most to climate change, and on a per capita basis, continue to contribute the most;
2 the industrialized nations have the greatest ability to pay for climate mitigation and adaptation;
3 the industrialized nations have benefitted the most from GHG emissions, while less-developed nations are bearing (and will continue to bear) a large portion of climate impacts. (see Shue 2001, p. 457)

It is a great irony that some of the worst effects of climate change—such as exacerbated drought in Africa or inundation of Bangladesh—are predicted to impact those parts of the world that have historically contributed the least to the problem. What's more, in the poorest of nations, there may be no way for people to achieve a minimally decent standard of living without actually *increasing* their GHG emissions—at least until infrastructure and technology that does not depend on fossil fuels can be established (see Shue 2001).

Whether one looks at historical GHG emissions, who has the greatest capacity to pay, or who has benefitted from the use of fossil fuels, a variety of different principles of equity "all converge on the same practical conclusion: whatever needs to be done by wealthy industrialized states or by poor nonindustrialized states . . . about global warming, the costs should initially be borne by wealthy industrialized states" (Shue 2010, p. 111). Of course, even if this is so, we face further questions about *how to institutionalize a just climate regime*. As mentioned earlier, one possibility would be to establish equal per capita emissions rights. An emissions allocation for each nation could then be determined based on this cap. (A nation's total allowable emissions would be calculated by multiplying the per capita allocation by the country's total population.) Although this would not genuinely hold wealthier nations responsible for past contributions to climate change, it would require wealthier and more industrialized nations to make significant cuts in emissions while allowing some room for further development by the lowest emitting nations. Such an approach may thus be "fair enough" (Baer et al. 2010): it may strike a reasonable balance between the genuine justice concerns of those who have contributed least to climate change, will bear the greatest burden, and have the least ability to shoulder that burden, on the one hand, and the inclinations of wealthy and powerful nations to resist responsibility for mitigation, on the other. In other words, an equal per capita approach may be the fairest of the politically feasible alternatives (Baer et al. 2010, p. 220).

In implementing this approach, one might determine the size of the per capita emissions rights (and hence the national allocations) in two ways:

1. One could ask what level of emissions would be sufficient for a person to live a minimally decent life,[1] then multiply this across the total population to determine the global cap on emissions, along with each nation's share of that cap; or
2. One could ask what level of global emissions would be sustainable, that is, what level of emissions the atmosphere could absorb without dangerously altering the earth's climate, then divide this by the total global population to determine allowable per capita emissions. National caps would then be determined accordingly.

The trouble is that although each of these strategies reflects a legitimate moral concern—the concern to allow each person a decent life, in the first case, and the concern to stabilize emissions at a level that will not endanger future generations, in the second—it may not be possible to address *both* concerns fully under current circumstances (see Shue 2001). In order to avoid excessive suffering in the near-term, option one therefore seems a better starting point. As Henry Shue (2001, p. 455) puts it, "to make the political choice to impose a ceiling on total emissions, while not guaranteeing a minimum to each person, would condemn to death the poorest people on the planet." However, a solution that takes seriously the looming impacts of climate change will need to reduce the value of individual emissions rights to create a declining cap in total global emissions over time. Such a solution would take option one as a starting point but gradually move toward option two.

To make an equal per capita emissions proposal more politically palatable, some have suggested a "contraction and convergence" approach, where equal per capita emissions are not required immediately, but instead are phased in. In this approach, the value of per capita emissions rights would increase gradually in less-developed countries and decrease gradually in more developed countries, to converge at equal (and sustainable) per capita levels. Such a scheme also could allow emissions trading, so that countries that find it difficult to meet their allowable cap could purchase emissions rights from those with unused emissions credits.

Although various strategies have been proposed in international negotiations, wealthy and poor nations remain largely at odds with one another regarding the distribution of climate responsibility. What's more, even establishing a just system for dividing responsibility *among* nations would not fully address the question of how to divide responsibility *within* nations. The division between "rich" and "poor" countries obscures deep inequities within nations, and the establishment of national targets, even if based on equal per capita emissions, does not ensure that each individual citizen actually *receives* the same right to emit. International climate negotiations typically treat nations as "black boxes" with respect to climate justice, and do not look within national borders to ensure an equitable distribution among the citizens of a particular nation.

Given the strength of national sovereignty, this is a difficult problem to overcome. However, it does illustrate the importance of addressing questions of climate justice at multiple scales. Climate policy needs to consider how GHG emissions vary within nations, not only between them. Paul Harris (2008), for example, has argued that wealthy people in relatively poor nations have obligations to reduce their carbon emissions even if their country's emissions as a whole are relatively low.

With respect to adaptation, differences across various temporal and spatial scales are equally important. Adaptation requires attention to local contexts and cannot be adequately managed from a global scale perspective (Adger 2001). However, adaptation cannot focus exclusively on the local or short-term time scales, because local and short-term adaptations can have effects that spill over to other locations or to future generations. While greater use of air conditioning may save people from illness or death during intense heat waves, the associated increase in energy consumption adds further to global warming in the long run (Adger et al. 2005). Similarly, fortifying a coastline against storms in one area may transfer impacts elsewhere (Pethick and Crooks 2000, Adger et al. 2005).

In finding just and politically practicable solutions to climate change, many issues remain. The relatively broad philosophical consensus that wealthy and industrialized nations should lead the way masks deeper disagreements over the fundamental principles that should guide climate mitigation and adaptation policies and how these principles should be institutionalized. In the case of mitigation, the equal per capita emissions strategy provides an ethically appealing option, but even if international agreement were to coalesce around that basic approach, further questions would need to be settled: the size of per capita emissions rights; whether emissions rights should be tradable, and if so, how; and how quickly the value of these allocations should decline over time to reach globally sustainable emissions levels.

Agreement on adaptation may be more difficult still because here there is no obvious policy "attractor" analogous to the equal per capita approach. The effects of climate change will continue to vary significantly throughout the world, and appropriate adaptation strategies will depend on the local effects of climate change as well as the social, political, and economic contexts in which these effects occur. The United Nations has established an Adaptation Fund to support adaptation in poor nations, and some wealthy nations have begun to contribute. However, as yet there is no philosophical or political consensus on who should pay for adaptation or how adaptation funds should be distributed.

The challenges of distributive justice in relation to climate change raise other issues that have received relatively little attention in either the academic or the policy realm. In particular, questions of *procedural justice* have largely been overlooked. Should policymakers and academic elites be the ones to decide what counts as a just distribution of responsibility for mitigation and adaptation? Given the heterogeneity in the effects of climate change both internationally and intra-nationally, can heads of state adequately represent the interests of their citizens in climate negotiations? If not, then what decision-making processes would be just? How might the perspectives of those with less power and wealth be best represented?

These questions draw attention to the fact that even the science of climate change is necessarily selective. Philosopher of science Nancy Tuana (2013), for example, argues that climate models and graphical representations of climate impacts capture only certain aspects of climate change, and that more needs to be done to take into account the gendered dimensions of global warming. Relatedly, Wylie Carr (personal communication) recounts the following debate between an African and a European scientist at a conference on geoengineering:

> [At the beginning of the conference], a scientific primer was provided for participants not familiar with the topic. The presenter, a natural scientist from the UK, provided model outputs in the form of four maps as key evidence for why he and his contemporaries in the United States, Canada, and Europe were taking certain geoengineering proposal[s] seriously . . . During this presentation, one of the few participants from a developing nation pointed out that the maps only looked at two parameters, surface temperature and precipitation. This was concerning to him because many of the people in his country in Africa depended upon fishing for their livelihoods and these models were not taking into account changes in ocean temperature or currents that could be detrimental to such livelihoods. The response from the presenter was that yes, there are many uncertainties that still needed to be examined by scientists and more models that need to be developed. In short, the conventional scientific response to concerns about uncertainty presented above. However, this individual from Africa, himself a climate expert who had served on the IPCC for 20 years, said

his concern was not about the model itself but the assumptions and ethics behind it.

This case seems to present a problem of *recognition*. Although scientists from a variety of countries and backgrounds were represented at the conference, in this case at least, the concerns of the African scientists were neither fully understood nor taken up by the presenter. Thus, as explained in Chapter 4, procedural justice is not merely a matter of formal representation but requires a process that enables genuinely diverse perspectives to be heard and taken into account.

Problems of justice are even more acute for future generations, for future people cannot participate in climate negotiations and it is easy for us to discount their interests. As debates over climate cost–benefit analysis illustrate, we lack consensus on how to treat future generations ethically or how to weigh the interests of the current generation in relation to those in the future. Philosophical frameworks for intergenerational justice remain inchoate, and have been hampered by the nonidentity problem, as discussed in Chapter 4. Some—such as Bjorn Lomborg (2001)—argue that we should not worry too much about future generations because they will be wealthier than we are. On this logic, sacrifices made by present generations to benefit future generations are seen as a transfer of wealth from those who are less well off to those who are better off. However, it is not clear that we can extrapolate the historic trend of increasing wealth indefinitely into the future, given the finite resources of the earth and the current trajectory of global climate change.

Finally, climate change raises questions of what we owe, ethically, to animals, plants, and other living things. The fate of nonhuman species gets almost no attention in climate policy debates, except insofar as these species serve as resources for human use. Similarly, writings on climate ethics often ignore altogether the possibility of ethical obligations that extend beyond the human community. In one prominent anthology on climate ethics, not a single chapter addresses the possibility of a nonanthropocentric climate ethic; another recent anthology contains just one chapter on the topic (Nolt 2011).

John Nolt (2011) argues that development of a nonanthropocentric, long-term ethics of climate change is critical. Only such an ethic, he believes, can contend with the implications of the harm we are causing in precipitating the sixth mass extinction since life on earth began. And only such an ethic can directly account for the harm climate change is causing and will cause to nonhuman living things. Many of the actions we take to address climate change for our own sake will also benefit nonhuman life; however, a nonanthropocentric climate ethic would diverge from an anthropocentric one in favoring mitigation (reducing GHG concentrations in the atmosphere to reduce climate change) over adaptation (adjusting to the effects of climate change) (Nolt 2011). Adaptation focuses primarily, if not exclusively, on human beings and communities; in contrast, by slowing the rate of climate change, mitigation benefits not only human beings, but animals, plants, and ecosystems more generally.

Perhaps, one reason why nonanthropocentric ethics have gained so little attention in climate ethics debates is because it has been difficult even to convince policymakers to take seriously climate justice issues involving present and future *human beings*. From a practical perspective, developing any kind of robust mitigation policy would be a tremendous victory. In the ethical triage of the climate policy arena, humans tend to win out. Yet, if we really do have obligations to nonhumans—as seems plausible, at minimum in relation to other sentient beings—then nonhumans should figure in our deliberations about how to develop a just and ethical climate regime.

In the end, despite growing philosophical consensus on certain key features of a just climate regime, more work is needed to develop both theoretical and practical frameworks that fully account for the procedural and substantive dimensions of justice for present and future people, and to incorporate concern for nonhuman species and ecological systems as a whole. Even where there is agreement on basic principles, institutionalization of these principles poses a significant challenge. On the positive side, however, climate justice is an area of great possibility, precisely because many questions remain unsettled. This is an area where philosophers can fruitfully partner with others (economists, social scientists, and policy specialists) to develop creative ideas that combine ethical principles with institutional frameworks that embody them.

For further thought

1. What key questions of justice does climate change pose?
2. Would an equal per capita emissions approach to setting national emissions limits be fair? Why or why not?
3. In what ways are procedural justice and recognition important to climate justice?

Geoengineering

Until recently, much of the discussion of climate ethics focused on mitigation and adaptation. Geoengineering is the new kid on the block. Although efforts to influence the weather have been utilized sporadically for decades, proposals to engineer the earth's climate on a grand scale have just begun to gain serious attention. In 2009, the Royal Society published a report on geoengineering that has been widely read. This report defines geoengineering as *"deliberate large-scale manipulation of the planetary environment to counteract anthropogenic climate change"* (Royal Society 2009, p. 1, emphasis in original). It then goes on to distinguish two main approaches in climate engineering—*solar radiation management* (SRM) and *carbon dioxide removal* (CDR). As the name suggests, CDR involves taking carbon dioxide out of the atmosphere to reduce the accumulation of GHGs. SRM, in contrast, aims to change the earth's energy balance by intercepting sunlight or reflecting it back into space, reducing the overall quantity of solar radiation received.

CDR and SRM are often discussed separately because they work in fundamentally different ways. CDR can include planting trees to take up additional carbon dioxide, enhancing carbon absorptive chemical weathering processes, or fertilizing the oceans to generate phytoplankton blooms that suck up CO_2. Proposed SRM techniques involve installing mirrors in outer space to deflect radiation away from the earth, painting rooftops or other surfaces white to increase the reflectance of the earth's surface, or perhaps most promising, injecting sulfate aerosols into the stratosphere to scatter and block incoming radiation.

As a general strategy, CDR seems to pose fewer ethical concerns than SRM. This is because CDR more directly addresses the root cause of the climate problem; in contrast, SRM seems to mask it. Rather than control GHGs, SRM aims to alter "something as fundamental as the amount of energy the earth receives from the sun" (Preston 2012, p. 3). As it turns out, it may be cheaper, faster, and technologically easier to counteract climate change through SRM than through CDR. Although the discussion below focuses primarily on SRM, it is important to recognize that certain CDR strategies could alter ecological systems significantly, and on a massive scale. Fertilizing the ocean with iron or other minerals in order to enhance CO_2 uptake provides one example. Ocean fertilization would alter ocean chemistry, and it could create anoxic "dead zones" and significantly alter marine ecosystems. Other CDR strategies such as planting trees to take up carbon dioxide raise fewer ethical concerns, but have weaker effects on global GHG concentrations.

Returning to SRM, a number of key ethical issues emerge. First, we face the general question of *whether it is ethically permissible to manipulate the climate intentionally at a global scale* (Jamieson 1996). Objectors describe geoengineering as "playing God," and as the height of human hubris. On the other side, advocates for geoengineering argue that humans are already exerting massive influence over the earth's climate: so why not exert that influence intentionally, rather than haphazardly, as we do through the burning of fossil fuels? This debate, in turn, raises questions about whether we have greater responsibility for *intentionally* changing the climate than for altering climate as a side effect of other activities (Jamieson 1996).

In addition to these very fundamental issues, SRM poses further ethical questions. First, there are issues of *procedural justice and consent* (Corner and Pidgeon 2010; Preston 2012). Who will decide when and how to deploy SRM? What is required for such a decision to be ethically legitimate? What form(s) of consent are appropriate or necessary? Or, as some have put it, who gets to set the global thermostat? In this regard, philosopher Kyle Powys Whyte (2012) argues that much of the current discussion of geoengineering is problematic. This discussion often frames SRM as "the lesser of two evils," suggesting a forced choice between geoengineering and catastrophic climate change. Yet this distracts from and dangerously sidesteps issues of consent. Questions of who participates and who decides whether to geoengineer are critical, especially since the people who have contributed the least to climate change and are likely to bear the greatest brunt of its impacts are the very same people who are in danger of having the least voice in decisions about geoengineering.

Another ethical issue has to do with broader questions of *global governance*. What sorts of laws should govern research and implementation of SRM? What recourse might a particular country have if SRM leads to unanticipated damage in its region? How can the international community prevent particular states (or individuals) from engaging in "rogue" geoengineering, or in attempting to use geoengineering to their own strategic advantage? These concerns are serious, and pursuing SRM may generate new sources of international conflict, which may in turn undermine the solidarity and sense of common purpose needed for productive international cooperation in response to climate change (Hourdequin 2012).

Relatedly, geoengineering raises concerns about what is known as a *moral hazard*. Situations of moral hazard arise when the costs of taking a risk are not directly borne by the individual, group, or institution that acts in a risky way. If you have insurance against the theft of your valuables, for example, you may be less cautious about protecting them. In the climate case, the idea is that geoengineering provides a kind of "insurance" against the worst effects of unmitigated GHG emissions—and thus may tempt us to avoid the hard work of mitigation.

Finally, there are questions about *uncertainty and risk*. How well can we anticipate the effects of geoengineering on the climate system as a whole? If we inject sulfate aerosols into the stratosphere, what unintended side effects might that have? Early models suggest that sulfate aerosol SRM may exacerbate the ozone hole, and disrupt Asian monsoons (Royal Society 2009). In addition, sulfate aerosol SRM may alter cloud-forming processes, which could accentuate or reduce its cooling effect (Kuebbeler, Lohmann, and Feichter 2012). Moreover, SRM—unless accompanied by aggressive mitigation—may lock us into perpetual geoengineering. If SRM occurred while GHGs continued to accumulate in the atmosphere, cessation of geoengineering would likely result in rapid and catastrophic climate shifts. Given the risks and uncertainties associated with geoengineering, how best can we decide whether to pursue it or not? In general, "[i]t is all but impossible in the early stages of a technology's development to know how it will turn out in its final form" (Rayner et al. 2013), and this is particularly true of geoengineering. What is more, geoengineering cannot be easily tested in a laboratory. Any large-scale test of geoengineering techniques will itself constitute a form of geoengineering. Thus, the line between research and implementation, in the geoengineering case, is blurred.

Currently, the debate over geoengineering is highly polarized. Some see geoengineering as a crucial insurance policy that can protect us from our own inability to address the root causes of the climate problem. Others see geoengineering as a "techno-fix" and an excuse not to address those very same root causes. In evaluating geoengineering, there are many serious ethical concerns to take into account. The debate over geoengineering involves technical questions, but it would be a mistake to see geoengineering merely as a technical problem. The risks associated with geoengineering are not only technical, but political, social, and moral.

The scale of geoengineering, the uncertainties associated with it, and the risk that geoengineering will further burden the least advantaged are all serious concerns. Issues of justice and recognition of diverse voices are very much at stake. To pursue geoengineering without involvement from all nations as well as from indigenous peoples would not only be unjust, but could undermine the fragile solidarity needed to protect human lives and cultures and to sustain the ecological systems upon which we depend (Hourdequin 2012).

For further thought

1. Is there an ethical difference between intentionally changing the climate through geoengineering and unintentionally changing by burning fossil fuels for energy?
2. What are the key differences between carbon dioxide removal and solar radiation management, from an ethical perspective?
3. Describe at least three ethical concerns associated with solar radiation management. In your view, which of these concerns is most serious?

Further reading

Adger, W. N. (2001). "Scales of governance and environmental justice for adaptation and mitigation of climate change." *Journal of International Development,* 13(7), 921–931.

Diffenbaugh, N. S. and Field, C. B. (2013). "Changes in ecologically critical terrestrial climate conditions." *Science,* 341(6145), 486–492.

Gardiner, S. (2006). "A perfect moral storm: Climate change, intergenerational ethics and the problem of moral corruption." *Environmental Values,* 15(3), 397–413.

Gardiner, S. M., Caney, S., Jamieson, D., and Shue, H. (eds). (2010). *Climate Ethics: Essential Readings.* New York: Oxford University Press.

Hourdequin, M. (2010). "Climate, collective action and individual ethical obligations." *Environmental Values,* 19(4), 443–464.

Hulme, M. (2009). *Why We Disagree about Climate Change: Understanding Controversy, Inaction and Opportunity.* New York: Cambridge University Press.

Jamieson, D. (1996). "Ethics and intentional climate change." *Climatic Change,* 33(3), 323–336.

— (2010). "Climate change, responsibility, and justice." *Science and Engineering Ethics,* 16, 441–445.

Johnson, B. (2003). "Ethical obligations in a tragedy of the commons." *Environmental Values,* 12(3), 271–287.

Preston, C. J. (ed.). (2012). *Engineering the Climate: The Ethics of Solar Radiation Management.* Lanham, MD: Lexington Books.

The Royal Society. (2009). *Geoengineering the Climate: Science, Governance, and Uncertainty.* London: The Royal Society.

Sinnott-Armstrong, W. (2005). "It's not *my* fault: Global warming and individual moral obligations," in W. Sinnott-Armstrong and R. Howarth (eds), *Perspectives on Climate Change: Science, Economics, Politics, Ethics.* Amsterdam: Elsevier, pp. 285–307.

Further reading

7

Ecological restoration

Introduction

Although climate change often appears as a daunting, large-scale environmental problem that exemplifies a troubled relationship between humans and the natural world, ecological restoration seems to offer the opportunity for redemption and a model for a more positive, constructive role for humans in nature. For centuries, humans have modified and manipulated their environments in order to make them more hospitable, to enhance soil fertility, or to sustain fish and wildlife species for food. Yet, over the course of the twentieth century, concerted efforts to restore entire ecosystems—not only for our sake, but for theirs—emerged. Restorationists began working to undo the ecological damage and degradation caused by logging, overgrazing, mining, and intensive farming. Restoration projects may be small—focused on establishing riparian vegetation in areas overgrazed by cattle—or large, aiming to restore entire watersheds or landscapes. Restoration is happening in rural areas, in wilderness, and in cities, and it is engaging people. For example, a local restoration organization in Colorado Springs, Colorado involves over 2,000 volunteers annually in their projects to reduce erosion, restore watersheds, and revegetate damaged areas following recreational impacts, wildfire, and other disturbances. Restoration, it seems, has the potential to heal both people and the land.

Perhaps surprisingly, some environmental philosophers have challenged this optimistic vision, arguing that ecological restoration "fakes nature" (Elliot 1982) or represents yet another way of dominating the natural world (Katz 1997). This chapter examines the possibilities and potential of restoration, while remaining aware of its limits. Ecological restoration is a particularly rich area to explore because it brings into focus many fundamental questions in environmental philosophy while at the same time revealing the very practical relevance of these questions. Is "nature" as it existed before human intervention something of special and distinctive value? Can we ever restore a damaged ecosystem to a truly natural condition? To what extent should social and cultural values influence the goals of ecological restoration projects? To what degree should restoration be guided by human needs? Is it ever permissible—or even desirable—to leave some remnants of past damage, as reminders of the effects of industrial use, military activity, or large-scale resource extraction? Given the dynamics of global environmental change, does it ever make sense to try to bring back ecological conditions that historically prevailed, or should we be engaging in forms of ecosystem reconstruction that look to the future rather than the past?

Land managers, local communities, restoration ecologists, government agencies, and others grapple with many of these issues as they face the very real challenge of setting restoration goals. The following section discusses the traditional goals and foundational ideals of ecological restoration and introduces challenges to these traditional ideals. The next section addresses philosophical concerns regarding the fundamental idea of ecological restoration. We then explore further challenges associated with complex socioecological landscapes and dynamic ecological contexts. Finally, we consider the role of restoration in mediating a more positive human–nature relationship, then take stock of ecological restoration and its potential in light of the many issues and questions raised here.

Ecological restoration: History, foundations, and fundamental goals

Stories of ecological restoration can be inspiring. Restoration projects can be big or small, and they can be developed and implemented by scientists, technicians, and engineers, or through community partnerships, volunteers, and local ecological knowledge. In many cases, volunteers work side by side with ecologists, hydrologists, wildlife biologists, and land managers to accomplish restoration goals. Although the restoration of degraded landscapes has a history many centuries long (Hall 2005), American conservationist Aldo Leopold is often cited as a critical figure in the development of contemporary ecological restoration. Leopold worked painstakingly to restore the Curtis Prairie at the University of Wisconsin Arboretum, beginning in 1930s, and in his spare time, he and his family worked to restore an old farm along the Wisconsin river, planting pines, aspens, and prairie flowers (Leopold 2004).

Today, restoration is happening all over the world. In the Netherlands, the Dutch are working to restore and rewild their landscapes by introducing

Konik horses and Heck cattle—proxies for extinct grazing animals—along the floodplain of the Rhine river, close to the German border. In the United States, large dams in Oregon, Washington, and New Hampshire have been removed to restore fish habitat and enable migration, and in northern Patagonia, Argentina, national park staff have worked to replace tree plantations with native forest. From planting trees to restoring streams to reintroducing animals, restoration is thriving, even as habitat destruction and ecological damage continue to generate the need for repair.

To understand the significance of ecological restoration—ecologically, socially, politically, and philosophically—we need to have some idea of what restoration *is*. Defining ecological restoration is more complex and controversial than it might seem. Since its founding in 1987, the Society for Ecological Restoration (SER)—the main professional organization for restoration ecologists and practitioners—has revised its definition of the term many times. In 1990, the organization set forth the following definition:

> Ecological restoration is the process of intentionally altering a site to establish a defined, indigenous, historic ecosystem. The goal of this process is to emulate the structure, function, diversity and dynamics of the specified ecosystem. (Higgs 2003, p. 107)

After numerous revisions and refinements, SER developed a description in 2002 that has survived more than a decade:

> Ecological restoration is the process of assisting the recovery of an ecosystem that has been degraded, damaged, or destroyed.

From a philosophical perspective, the shift from 1990 to 2002 is instructive. In comparing these two characterizations, the specificity of the first stands out in contrast to the breadth of the second. But perhaps more importantly, note what has been eliminated in the later definition. We no longer find the phrase "defined, indigenous, historic ecosystem"; instead, the newer definition emphasizes the idea of helping an ecosystem recover after degradation or damage, without specific reference to the past.

Although SER has settled on a definition of restoration for now, there are ongoing debates in ecological restoration surrounding the role and importance of history. Some restorationists place great weight on history, while others focus more on the idea of ecological health. Yet history seems implicit in the very idea of restoration. The term "restoration" suggests returning something—whether an ecosystem, an artwork, or an historic building—to some prior state. Architectural restoration brings back the distinctive historical qualities and characteristics of a building, while art restoration typically aims to restore the original qualities of a painting or other work. Similarly, ecological restoration evokes the idea of bringing an ecosystem back to its original, undisturbed condition. SER's 1990 definition resonates with this idea, marking the original state for a particular site as the native, historic ecosystem, or the ecosystem in its natural state.

In North America, the idea of restoring ecosystems to their historic natural condition has been highly influential, and many restorationists conceive of natural conditions as those that prevailed prior to European colonization, often referred to as *presettlement conditions*. However, the idea of presettlement conditions has been vigorously challenged by environmental historians, who have pointed out that America was neither unsettled nor pristine prior to Europeans' arrival. In fact, Native Americans not only lived throughout North America, but significantly altered ecological systems through hunting, clearing forests, burning prairies, and other practices (see Cronon 1995; Denevan 1992). The very idea of presettlement conditions in the United States and Canada, when used as shorthand for conditions prior to European settlement, seems to deny either the presence or significance of Native Americans prior to colonization. Given what we know about the history of North America and its peoples, the notion of pre-European pristine nature is difficult to defend.

It is possible, however, to reject the North American presettlement ideal and to remain focused on returning ecosystems to their *natural condition* as the key goal of restoration. Typically, the "natural" is taken to exclude the human, implying that a natural ecosystem is one undisturbed by humans. Thus, ecosystem conditions prior to human disturbance set the baseline for restoration. Since restoration cannot reverse time, restoration of degraded land can never fully succeed in restoring naturalness: it can never transform a disturbed ecosystem into a fully undisturbed one. Nevertheless, restoration may be able to return an ecosystem to something *approximating* its natural state. In theory, at least, restoration can bring back the plant and animal species that prevailed prior to disturbance, generating an ecosystem with similar structure and function to that which came before humans intervened. Many restorations do, in fact, aim at something along these lines.

The idea of predisturbance conditions can also be adapted to loosen the connection with naturalness. Rather than focus on bringing back conditions prior to all human influence, we can think of restoration operating at different scales in space and time, depending on the focal disturbance in question. Restoration of a stream damaged by grazing might seek to emulate the channel structure, bank structure, and plant composition that existed prior to the introduction of domestic cattle. Such restoration might focus on a relatively small stretch of heavily impacted stream or extend to an entire watershed. Wildlife restoration in a national park like Yellowstone might aim to bring back animals that were extirpated by hunting many decades prior.

The reintroduction of wolves to Yellowstone in 1988 aimed at just such a goal.

By now, it should be clear that the question of how to understand the goals of ecological restoration is not merely scientific, but at the same time philosophical, social, and political. Nevertheless, science plays a crucial role in the practice of ecological restoration, including the establishment of restoration goals. In order to restore a forest that has been heavily logged, or whose composition has been significantly altered over time through the suppression of natural fires, we need to know what kinds of trees and plants grew there historically and what conditions will enable these species to thrive. Historical ecologists help reconstruct the ecological past using a variety of methods, including tree ring analysis, pollen records, and reconstruction of fire histories using burn scars on trees. Ecologists also study the habitat requirements of various animals, the environmental conditions necessary for the growth of particular plants, and the way in which various species interact to compose ecological communities. Other scientists—such as hydrologists and soil scientists—help us understand the structure and dynamics of rivers or the cycling of nutrients through the soil.

Science itself, though critical in identifying reference conditions for restoration, reveals complications associated with the goal of returning an ecosystem to a particular predisturbance state. Early twentieth century ecologists generally believed that ecological systems undergo defined patterns of development until they reach an equilibrium, "climax" state, and drawing on this view, one might think that restoration should aim to reproduce this climax condition. By the 1980s, however, a new ecological paradigm was emerging, in which ongoing disturbances—through fire, wind, flood, or waves—were understood to be a central part of virtually all ecological systems. Thus, using a snapshot in time or the idea of a stable, climax state as the basis for establishing restoration goals can overlook natural dynamics, by privileging stability over change.

In response to these developments, ecologists developed the concept of *historic range of variability*, which helps capture the way in which ecosystem conditions fluctuate over time in response to typical patterns of disturbance (Keane et al. 2009). The age and composition of trees in a particular forest ecosystem, for example, might vary over time in response to recurring lightning-ignited fires. Historical ecologists can reconstruct both fire patterns and shifts in forest structure and composition through a number of decades. Restoration ecologists can then broaden their targets accordingly, focusing less on replicating the precise structure and composition of an ecosystem at a fixed historical point in time, and more on restoring processes that renew the system's historical dynamics. However, as we will discuss, large-scale global changes—and climatic changes, in particular—have called even this more dynamic understanding of restoration goals into question. Given significant projected shifts in global temperature, precipitation, and other climatic features, some wonder whether historical ecological conditions remain relevant at all.

There is one further complication related to restoration goals that is worth noting here. The development of the field of ecological restoration has been strongly shaped by North American restoration practices and examples, which have tended to emphasize the ideal of naturalness, where restoration aims to create ecosystems as they existed prior to significant human influence, or as they would exist today if humans had not significantly altered them. Yet naturalness, at least as traditionally understood, may be appropriate in some areas, but not others. Many restorationists outside of North America have hesitated to adopt restoration goals grounded in the "natural," insofar as this is taken to exclude humans. In Europe, there is deep awareness of thousands of years of continuous human occupation and land use. Walking near the border of the Netherlands and Germany, for example, one encounters Roman ruins. Along the Mediterranean, the visible remnants of human occupation are even more striking. The idea of a fully pristine or unoccupied European landscape is thus difficult even to imagine. Accordingly, restoration goals in these places must explicitly negotiate the complex and ongoing intermingling of humans and nature, an issue to which we return in the section "Hybrid landscapes, climate change, and other challenges".

For further thought

1. What are the limitations of using "presettlement conditions" as a baseline for restoration?
2. How might the concept of historic range of variability change traditional restoration goals?
3. Consider a familiar place in need of restoration. How would you go about setting restoration goals for that place?

Authenticity and historical fidelity in ecological restoration

As discussed above, the traditional goals of ecological restoration have centered on the idea of returning an ecosystem to its natural state, or its condition prior to human disturbance. This idea has been captured in two key guiding values for restoration—authenticity and historical fidelity—whose meanings are often intertwined. More precisely, the idea of historical fidelity often functions to give content to the broader idea of authenticity in ecological restoration. Authenticity is a general ideal for restoration, and is tied to ideas of genuineness and legitimacy. According to the *Oxford English Dictionary*, "authentic" can mean authoritative, valid, genuine, original, or real. Although references to the genuine and the original suggest a connection to the past, authenticity is perhaps best understood as a generic ideal, where *authentic restoration is restoration that embodies the central goals of restoration practice, however such goals are defined*. Thus, the ideal of authenticity gains its full meaning in relation to a fuller, more clearly specified purpose of restoration.

We have seen that one prominent view of restoration's central purpose involves *historical fidelity*, or faithfulness to the natural conditions that prevailed prior to human disturbance. From this perspective, authenticity and historical fidelity in ecological restoration are almost synonymous: an authentic, or genuine, restoration is one that faithfully returns a site to its historic natural state. Binding authenticity to history, this view echoes the idea of restoration as aiming to reproduce a "defined, indigenous, historic ecosystem," as the original SER definition suggested. Genuine restoration is that which returns a place to its original condition, or something closely approximating it. Just as an authentic architectural restoration preserves, recreates, and reveals the original style and materials of a building, an authentic ecological restoration recreates the original elements and structure of an ecological system.

Others think that authenticity can cleave from history, and that genuine restoration need not fixate on the past. Restorationist Andre Clewell (2000), for example, distinguishes between *natural authenticity* and *historical authenticity*. Historical authenticity in ecological restoration requires matching the restored ecosystem with a historical reference state, whereas natural authenticity emphasizes returning an ecosystem to a healthy condition, whether or not the ecosystem precisely reflects its historic structure and composition. For Clewell, natural authenticity is achieved when restoration produces an ecosystem capable of ongoing self-renewal, one that "self organizes through natural processes" (2000, p. 216). Natural authenticity, he suggests, is a more reasonable goal because it does not require precise knowledge of a site's ecological history, and it achieves the fundamental aim of restoration, which is "not to revive the past in order to indulge our nostalgia, but to secure our future by restocking a dangerously depleted global inventory of natural areas" (2000, p. 217). Notice that Clewell's understanding of the natural departs from the view described above. Rather than emphasizing conditions prior to human influence, Clewell's understanding of naturalness stresses the ability of a natural system to persist without ongoing human intervention, regardless of its resemblance to a prior undisturbed state.

Others take issue with the prospect of divorcing restoration from history. In particular, authors such as Eric Higgs (2003) and Dave Egan (2006) worry that history is what anchors restoration, and that without it, restoration will become unmoored, driven by parochial and capricious human interests. According to this view, restoration without history is no restoration at all. Although we may be able to design and create healthy, novel ecosystems in formerly degraded sites, unless these new systems bear some resemblance to their historical antecedents, they cannot rightly be called restorations. From this perspective, historical fidelity remains an important guiding value: restored ecosystems must be faithful to what came before.

The justification for historical fidelity is often tied to the idea that restoration is about healing nature and not primarily about satisfying human desires:

> Paying attention to the historical fidelity or genuineness of our projects can help greatly to move us [toward an appropriate relationship with nature] by forcing us to look deeply into those beings and forces we seek to unleash so that they may pursue their way of life. Otherwise, we are likely to fall prey to the mass consumerism that surrounds us—creating gardens where we maintain beings as "things" strictly for our use and admiration. (Egan 2006, p. 224)

William Jordan and George Lubick (2011, pp. 2–4) echo these ideas, arguing that ecocentric restoration—which is "focused on the literal re-creation of a previously existing ecosystem"—has value because it forces us to subordinate our own interests and to encounter nature as something other than and independent of us.

Although these rationales emphasize an ecocentric perspective, a role for history can also be justified in relation to human values and sources of meaning. Eric Higgs (2003, pp. 143–145) argues that the impulse to restore is driven, in part, by a nostalgic desire to return to an earlier, simpler time. History is also important because it allows us to understand restoration as part of an ongoing narrative of a place, in which we work to make sense of the future in light of the past (Higgs 2003, pp. 145–154). Finally, Higgs emphasizes the role of history in giving us a sense of what he calls "time depth": just as we prize buildings that endure through the centuries, serving witness to social and cultural changes across generations, we value ecosystems that endure, preserving a sense of continuity amidst the dynamics of ecological and evolutionary changes (p. 155). Preserving a sense of continuity with history is thus an important function of restoration. Ultimately, Higgs argues that ecological restoration derives its meaning for human beings from its connections to history, and for this reason, history must remain an important part of restoration practice. Nevertheless, it is not *only* history that matters.

Higgs believes that twin goals of ecological integrity—which emphasizes the health and wholeness of ecological systems—and historical fidelity are at the core of authentic ecological restoration.

The debate about how to understand authentic ecological restoration and its relation to history is ongoing, and despite disagreements, parties to the debate—ecologists, restoration practitioners, and others—generally agree that genuine ecological restoration is possible, and that restoration, in general, is a good thing. Restoration projects take degraded, depauperate landscapes and transform them into functional, diverse ones, and there is something deeply satisfying about transforming a barren clearcut into a space full of tree seedlings, or seeing salmon return to a newly undammed river. Some philosophers, however, have expressed deep skepticism about restoration, questioning its value and the very possibility of authentic restoration.

For further thought

1. On what grounds might you argue that historical authenticity, or historical fidelity, should be the goal of ecological restoration? On what grounds might you argue for what Clewell calls natural authenticity?
2. Is history more important to restoration in some contexts than others? Why?

Faking nature?

For Robert Elliot (1982), all restoration "fakes nature." Ecological restoration overreaches by promising something it can never deliver: a natural landscape. Thus, restoration is *never* authentic and always falls short of its aim. Elliot introduces his argument with an example. Imagine a proposal to mine beach sands for the mineral rutile. The miners insist that their project should be approved, arguing that they will restore the beach afterwards, leaving everything just as it was before the disturbance. There is no reason to reject the mining proposal, they claim, as restoration will fully return the natural value of the site. Elliot calls this idea—that restoration can fully restore natural value—*the restoration thesis*. He vigorously denies it. At the heart of this denial is the idea that ecological damage by humans creates a disruption and loss of value that can never be fully repaired, *even if the original ecosystem is perfectly replicated*.

At first, this idea may seem counterintuitive. If the restored ecosystem is just like the original, how could it be less valuable? In reply, Elliot calls our attention to the way in which the genealogy, or history, of a thing affects the way we value it. We can adapt one of Elliot's examples to illustrate the point. Imagine that your grandmother is an artist and leaves you one of her most beautiful paintings when she dies, which you keep in a treasured location at home. Unbeknownst to you, an admirer of her work sneaks into your home and steals the painting, replacing it with a replica. The replica is excellent, and you would never notice the difference. Nevertheless, if you were to find out that the original had been replaced, you would feel a deep sense of loss, and you would regard the replica as a fake, of radically lower value than the original. Similarly, argues Elliot, an ecosystem that has been destroyed then put back together is not the same as the original. It is a replica and a fake, and has less value than the original. Just as the value of your grandmother's painting derives, in part, from the fact that she was the one who painted it, the value of a natural ecosystem derives, in part, from the fact that it was nature that generated it. A restored ecosystem can have value, but never a value equal to the natural system it emulates. Human disturbance disrupts the natural genealogy of a place, compromising the continuity of its history.

It is worth pausing at this point to consider whether you agree with Elliot's analysis. Is all restored nature "fake" in some sense? If so, does this mean that restoration is not worth pursuing?

There seem to be two senses in which Elliot suggests restoration fakes nature, an ontological sense and an epistemic sense (Hourdequin and Havlick 2013). (Recall that the term "ontological" refers to the nature of being, and "epistemic" refers to the nature of knowledge.) First, restored landscapes are fake in the sense that they are not original, genuine, natural landscapes. This is the ontological sense of "fake," since it focuses on the actual landscape, its characteristics and mode of existence. Second, restored landscapes are fake in the sense that they are deceptive: they try to pass themselves off as something they are not. This is the epistemic sense: restored landscapes disrupt knowledge and understanding of what is truly natural by falsely imitating the natural. It might be possible to engage in restoration without faking in this latter sense, without deceiving observers into thinking they were witnessing an undisturbed natural landscape. But if we accept Elliot's arguments about disrupted natural processes, then restored landscapes will always be ontologically different from their undisturbed counterparts, and in this sense, "fake."

Although Elliot is careful to explain that he does not oppose restoration per se—after all, a restored landscape may have more value than the degraded landscape it replaces—he does want to defuse the restoration thesis and the associated argument that we need not worry about damaging nature, for we can always put it back. Nevertheless, the faking nature argument casts a negative shadow on restoration more broadly: certainly, restoration volunteers are not typically motivated by the thought that they are working to create a false imitation of nature.

Many dispute the view that restoration fakes nature. Some question the analogy between art forgery and ecological restoration. Others wonder whether Elliot relies on too sharp a dichotomy between humans and nature, between the sullied and the pristine. Regardless, the faking nature argument poses a challenge to those who wish to explain and justify the distinctive value of restoration.

As we have seen, Elliot's arguments highlight the dangers of using the possibility of restoration to justify ongoing ecological damage. Philosopher Eric Katz (1997) has argued even more vehemently against ecological restoration, calling it "the big lie." Although we may be inclined to vilify logging, mining, and other forms of resource extraction while valorizing restoration, Katz believes that in both cases, humans dominate and oppress nature. Whether through logging or tree planting, we are shaping nature to conform to our will, rather than allowing a space for nature's autonomy. Building on Elliot's argument, Katz suggests that restoration cannot recreate a natural system, but instead generates a human artifact, an object produced with specific human purpose and intent. Because they serve human purposes, artifacts are fundamentally anthropocentric, according to Katz. A natural system, by contrast, is one that lacks a purpose imposed from outside. Instead, free from human domination, an autonomous natural system is one that "[pursues] its own independent course of development" (p. 105).

While Katz's position may seem disheartening, his arguments have been countered from multiple perspectives. For example, where Katz lumps all human manipulation and management of nature into a single category involving the imposition of anthropocentric goals, we might distinguish various forms of human engagement with nature. Although all human influence on nature can be understood as *anthropogenic*—human generated or of human origin—not all human influence is necessarily *anthropocentric*, or focused on narrow human interests. Following this line of thought, restoration might be viewed not as a mode of human domination but rather as a model of partnership between humans and nature, as providing an opportunity for humans to engage constructively as active ecological citizens (Light 2005). From another perspective, one might argue that restoration forces us to confront the ecological damage we have caused, and to grapple with the complexities of our status as human beings, who are at once part of and apart from nature (Jordan and Lubick 2011). As discussed below in the section "Restoration, participation, and engagement," this latter view makes space to acknowledge the harm that restoration itself can cause by removing some organisms in order to bring back others. Such a perspective allows for the possibility that restoration is an important and valuable practice, yet one fraught with moral ambiguity.

Because restoration takes so many different forms, a nuanced and context-sensitive approach to understanding and evaluating restoration practices may be most reasonable. Worries about faking nature and human domination through restoration remain relevant, but they take different forms in different contexts, and bear greater relevance to certain kinds of restoration than others.

For further thought

1 In your view, can humans create nature through restoration? Or is "human-created nature" an oxymoron?

2 Is restored nature always of lesser value than undisturbed nature?

3 How might you argue against Katz's view that restoration is just another form of human domination of nature?

Hybrid landscapes, climate change, and other challenges

Under a very simple model of restoration, there is a neat division between nature and humans, and between disturbed and undisturbed landscapes. Although a restored landscape can never be pristine, restoration aims to bring a site back to resemble an idealized natural condition, and to erase as fully as possible the marks of human influence. Of course, the division between humans and nature is not as neat as one might suppose, and many landscapes formerly considered "pristine" were the product of ongoing human influence. What's more, ecological systems are not static, even in the absence of any human influence at all, so aiming for a historic reference condition may not put an ecosystem back on its natural trajectory. Finally, human influence across the globe is now pervasive. As Bill McKibben (1989) argued more than two decades ago, through climate change, organic pollutants that have found their way to the far reaches of the Arctic, and numerous other means, humans have altered virtually the entire planet, leaving no place undisturbed. If we have truly reached the "end of nature," as McKibben suggests, what should restoration aim to restore?

These challenges raise important and difficult questions for restoration, yet it is important not to overstate the case. It is not as if restoration is impossible or incoherent without a sharp divide between humans and nature, and even in a human-influenced world, we should not be too quick to pronounce nature dead or gone. What the complications show, in part, is that our models for restoration have been too simple and our distinctions too stark. Geographers, historians, and sociologists have drawn our attention to this point by describing and exploring *hybrid landscapes*, landscapes that confound a binaristic nature/culture distinction (Whatmore 2002; White 2004). Hybrid landscapes are neither fully natural nor fully cultural, instead they mix and blend these categories in ways that prompt us to think more carefully about the values they harbor, as well as the complex interrelationships between humans and nature.

As ecofeminists have so clearly emphasized, binary distinctions tend to encourage neatly assigned moral valences, with one member of the binary given positive value, and the other associated with the negative. Historically, this has worked both ways with respect to the nature/culture distinction. The Western focus on dividing nature and culture has justified elevating culture over nature, as expressed in the goals of "civilizing savages" and "taming wilderness" among early European settlers in America (Nash 2014). This view represented human progress as a process of bringing nature under cultural control. The countervailing position retains the nature/culture distinction, yet valorizes nature. This view is represented in the writings of American wilderness advocate John Muir, who sees nature as redemptive, and civilization as obscuring what is of true value:

> In God's wildness lies the hope of the world—the great fresh unblighted, unredeemed wilderness. The galling harness of civilization drops off, and wounds heal ere we are aware. (Muir 2001, p. 315)

Yet neither of these binaristic views is complete. Civilization, or culture, is not just a "galling harness" nor is it free from flaws. Similarly, nature cannot be neatly categorized as either an unruly beast or an earthly paradise. Our views of nature, culture, and their interactions are inevitably value laden, but that need not force us to treat them as monolithic, opposing, and mutually exclusive.

By turning our attention to the ways in which nature and culture intermingle, we can see more clearly the complex interactions between them. Humans live in mixed landscapes: from wilderness to farms to suburban backyards to city parks, humans and the rest of the natural world are deeply intertwined. Take agricultural lands as an example. Farmers plant crops and they raise animals. The crops are plants derived from wild ancestors through centuries of selective breeding; the animals are domesticated, yet remain connected to the wild. How are we to categorize these denizens of agricultural life? Certainly cows, pigs, corn, and wheat are the products of culture, but aren't they at the same time, in some sense, natural? When our classifications are disrupted, we are forced to reckon with the nuanced ethical challenges of human relationships with nature rather than easily sort good from bad based on simple categories.

Two examples can help illustrate the potential fruitfulness of hybrid landscapes in embracing complexity and developing new models for ecological restoration. Consider ecological restoration in the Netherlands. The Dutch use the term "new nature" to refer to areas where former agricultural lands—or even lands reclaimed from the sea—are naturalized and restored (see Van Der Heijden 2005). Although Katz and Elliot might object, the Dutch call the work to restore such lands "nature development," and they have worked to remove dikes, allowing rivers to flood, and to introduce semi-wild grazing animals to take the place of extinct herbivores. One of the most unusual examples of "restoration" is found in the Oostvaardersplassen, an area of land reclaimed from the sea and originally intended for industrial development. After draining, an area of newly created wetlands of Flevopolder began to attract large numbers of breeding birds and the Dutch government eventually decided to protect the area as a park (Van Der Heijden 2005; Kolbert 2012). In the 1980s, Heck cattle and Konik horses were introduced, and these animals now roam freely throughout the park. A place such as Oostvaardersplassen confounds traditional categories. Though the Dutch Forestry Commission (Staatsbosbeheer) describes it as "nature at its most pure," Oostvaardersplassen is clearly not an ordinary nature reserve. The site raises questions about the ways in which humans can "create" nature and about the moral status of animals undergoing "de-domestication" (Gamborg et al. 2010). Such animals are neither fully domestic nor fully wild, but straddle two categories and challenge us to determine the nature and extent of our obligations toward them (Gamborg et al. 2010). In doing so, they call into question the adequacy of our restoration concepts and frameworks, and encourage us to imagine new ways of understanding restoration's possibilities.

One way of opening up possibilities for ecological restoration is to allow for—or even embrace—the intermingling of nature and culture. Such strategies are particularly appropriate for landscapes with significant political, social, and cultural histories, where nature and culture have co-constituted one another over time. For example, many former military sites in the United States have been converted to National Wildlife Refuges over the course of the last 25 years, and some of these lands are the sites of extensive restoration. Military-to-wildlife refuge conversions give these lands a new central purpose—the protection of wildlife and their habitat. At one refuge, the Rocky Mountain Arsenal National Wildlife Refuge near Denver, Colorado, cleanup of toxic contamination associated with chemical weapons production has been followed by extensive and ongoing efforts to restore native short grass and mixed grass prairie and to reintroduce bison, animals that roamed the plains at the foot of the Rocky Mountains in large numbers until the late 1800s. Visions for the restored site vary. While refuge managers and a majority of visitors favor a landscape that reflects ecological conditions prior to extensive European settlement, some members of the local community—particularly those committed to preserving its history—would like to see these traditional restoration goals augmented by integrating elements from the layered history of the site (Havlick et al. 2014).

Although there are financial and institutional obstacles to such integration, the juxtaposition of nature with relicts of past military or agricultural use intrigues visitors to other military-to-wildlife conversion sites, and could provide a way of retaining visible elements of past uses and the way in which humans and nature interact to shape landscapes over time. In some places, postindustrial sites have been transformed into new gardens and public spaces, while retaining visible elements of the past. Wildlife managers at some eastern U.S. refuges such as Assabet River and Great Bay National Wildlife Refuges are exploring the possibility of converting old concrete bunkers into bat hibernacula. These blendings of human histories with restoration and wildlife protection productively complicate our understandings of these sites—challenging overly dichotomous distinctions between humans and nature, while also encouraging us to confront the legacies of militarization, weapons production, and toxic contamination.

The need for restoration to grapple with the complexities of human–nature interrelationships is not confined to former military sites or lands reclaimed from the sea, however. Traditional restoration goals are further challenged by global climate change as well as land use changes that alter the landscape-scale contexts in which restoration takes place. Often, restoration sites are islands in larger landscapes of intensive human occupation and development, and smaller sites may be unable to support plant and animal communities that require larger continuous areas to persist. With respect to climate change, changes in temperature, precipitation and patterns of disturbance at the regional scale may generate conditions that can no longer support the species that existed there prior to human disturbance. Some scholars have gone so far as to suggest that traditional restoration goals are no longer relevant, and we need a new paradigm focused on future-oriented restoration (Choi 2007).

Defenders of future-oriented restoration argue that it is the most appropriate response to a rapidly changing world. Forward-looking restoration would:

1 Focus on restoration goals appropriate to future environments rather than past environments;

2. Accommodate change by allowing for multiple endpoints and trajectories for restored sites;
3. Emphasize ecological function rather than structure and composition; and
4. Embrace the role of values in ecological restoration, taking into account social and economic considerations. (Choi 2007)

The rationale for the first three goals draws on the point that ecosystems are changing rapidly—and restoring ecosystems to historic baselines may no longer be possible or sustainable. Instead, it may be more appropriate to focus on restoring healthy ecosystems that are resilient in the face of environmental change. If restoration is no longer tightly bound to historical fidelity, however, then there may be a variety of ways of achieving the general goal of creating healthy and resilient ecosystems. This is where values enter in: since restoration goals cannot be objectively determined, we need to take account of social, economic, and ethical values in setting restoration goals for a particular site.

Some have taken the response to changing environments a step further, suggesting that the framework of restoration should be subsumed under a broader umbrella known as "intervention ecology" (Hobbs et al. 2011). Intervention ecology calls attention to the ways in which we actively and intentionally intervene in ecosystems to maintain or alter them, and would encompass both restoration and conservation. Although there is some attractiveness to an approach that unifies restoration, conservation, and land management, it is reasonable to worry that in lumping all of these activities under the broad category of "intervention," important distinctions will be lost. What's more, as Hobbes et al. (2011, pp. 447–448) note:

> [T]he term intervention is itself loaded . . . It is certainly not the nurturing term that restoration is, and it is hardly likely to engage communities in ecosystem management in the way restoration does . . . Maybe this connotation alone indicates that it would be wrong to advocate doing away with the idea of restoration altogether and that we should instead accept that it is one particular type of intervention that fosters community engagement with nature.

Insofar as restoration loses sight of history altogether, it risks collapsing into a Promethean remaking of the world, as if the current and historical ecological relationships at a particular site have no meaning. Without some sense of what came before, we may fail to be receptive to the possibilities for and value of some degree of continuity over time. At worst, we may begin to see the natural world as providing a set of pieces and parts that we can rearrange at will, in order to optimize the satisfaction of our own desires and needs.

Although restoration surely must take into account the future and acknowledge ongoing environmental change, it is not clear that we need to move beyond restoration entirely, or to look only to the future in restoration, without ever turning to the past. In some instances, aiming to bring back a particular predisturbance ecosystem may be the best thing to do; in others, it may make sense to build on what already exists at a site, even if what exists now is a product of the joint interaction of humans and nature. Restoration is a context-sensitive practice, capable of encompassing a plurality of goals and rationales. The debates discussed above are valuable insofar as they make us more reflective about these goals and rationales, and sensitive to the ways in which a single, unified understanding of what restoration is for may serve neither us nor the natural world well.

For further thought

1. Why is the concept of the hybrid landscape important to restoration?
2. Should restoration be subsumed into a broader framework of intervention ecology?

Restoration, participation, and engagement

It is interesting to note that despite their tentative advocacy for "intervention ecology," Richard Hobbs and his colleagues suggest that the practice of restoration may have a particular value that "intervention" cannot capture (2011, pp. 447–448). Many who have engaged in or written about restoration share the view that restoration makes possible a valuable form of human engagement with nature. While environmentalists frequently decry humans' capacity to damage, destroy, and desecrate nature, restoration seems to offer a hopeful alternative. Restoration can serve as a form of reconciliation between humans and nature, or as restitution for environmental damage. Restoration may provide a model for ecologically engaged citizenship, and it may "rebuild our concern with things that matter" by "[nourishing] both nature and culture" (Higgs 2003, p. 226). Below, we examine the work of three scholars who have articulated in distinct ways the value of restoration for human communities and relationships with nature.

In response to critiques of restoration as "faking nature" or "the big lie," philosopher Andrew Light (2003) has argued that we can distinguish between two kinds of restoration, malicious and benevolent. Malicious restorations are those that help to rationalize ongoing ecological damage. Arguments that invoke the possibility of future restoration as a reason for ecological destruction support this kind of malicious restoration. Benevolent restorations, on the other hand, are "those undertaken to remedy a past harm to nature" without attempting to justify or rationalize such harm (Light 2003, p. 401). By making this distinction, we can acknowledge the risk that restoration may become yet another excuse for destroying nature, or another vehicle for domination, while at the same time opening up a space in which we can see the value that ecological restoration can have.

The value of good restoration is arguably twofold: first, it can repair ecological damage, restore ecological function, and renew the capacity of land and waters to support healthy plant and animal communities. Second, ecological restoration can engage us in constructive relationships with the natural world in which we acknowledge our destructive power and take responsibility for the damage we have done. This second aspect of restoration has been discussed from different perspectives.

Andrew Light (2006) suggests that ecological restoration can contribute to the development of a form of ecological citizenship that would bring us into better relationship with the natural world. The idea of ecological citizenship builds on a more foundational idea of ethical citizenship, where citizenship is not merely a legal category that confers certain status and associated rights. Instead, an *ethical* notion of citizenship requires constructive public engagement; one that helps strengthen the community of which one is a part. To this notion of citizenship—which is significantly more demanding than merely following the laws and voting in democratic elections—Light adds an ecological dimension. He explains:

> To add an environmental dimension to this expanded idea of citizenship would be to claim that the larger community to which the ethical citizen has obligations is inclusive of the local natural environment as well as other people. That is not to say that all legal citizens of a community would be required to become environmental advocates or ecological citizens in this way, but, rather, that embracing the ecological dimensions of citizenship would be one way of fulfilling one's larger obligations of this thicker conception of citizenship. (Light 2006, p. 177)

Public engagement in restoration work can play an important role in cultivating ecological citizenship, because it establishes connections among people, and between people and their local environments. The relationships people form with the natural world through on-the-ground restoration work are crucial, for "our relationship to nature is ultimately shaped locally" (Light 2006, p. 171), and a personal connection to the land supports forms of environmental stewardship that cannot be generated by regulations alone.

While Light emphasizes the connections between restoration and engaged citizenship, Eric Higgs (2003) suggests that participatory restoration can serve as an antidote to our technological culture and the alienation it produces. For Higgs, the participatory and grassroots dimensions of restoration risk being lost as restoration becomes increasingly professionalized. He worries that a science and engineering approach to restoration will result in the exclusion of local perspectives and knowledge, a stronger emphasis on prediction and control, and a loss of humility in restoration. This in turn may lead to the commodification of restoration, in which corporations and governments take over restoration in ways that emphasize expertise, efficiency, and uniformity at the expense of locality and particularity. *Technological restoration*, as Higgs calls it, moves restoration out of the community and into economies where restoration projects and technologies are bought and sold: they become commodities in a larger system of trade and consumption. Restoration—in this sense—loses its rootedness in place and the relationships that tie it to place. In contrast, *focal restoration* retains this connectedness, and while retaining a place for scientific knowledge, with its emphasis on generality and universality, focal restoration draws our attention to the *practice* of restoration and how it ties people to place, and to one another.

The idea of focal restoration draws heavily on the work of philosopher of society and culture Albert Borgmann. Borgmann (1984) argues that one of the critical losses associated with our technological culture has to do with the replacement of focal practices by what he calls the "device paradigm." Focal practices are those that connect us with one another and to place. By doing so, focal practices center and ground us, giving meaning to our lives. Focal practices might include gathering to make music with a group of friends, or preparing and enjoying a communal meal. The technologies involved in focal practices typically require skill to master and they themselves represent continuity and facilitate engagement. A musical instrument provides one example, a set of garden tools—a hoe, pitchfork, rake, and trowel—might be another. These technologies can be contrasted with "devices," which provide goods and services, but without engaging us in the process of producing those goods. A central heating system, argues Borgmann, is a device, while a woodstove is a focal thing, a technology that makes possible our ongoing engagement with others and with the natural world. The central heating system requires very minimal engagement: set the thermostat, and you are done. The woodstove, in contrast, engages us in the skillful practice of chopping and splitting wood, stacking it, and in this way involving ourselves directly in the heating of our homes.

Focal restoration engages us similarly. It is not a high-tech, low-involvement, technological solution to ecological degradation. Instead, focal restoration requires that we work with others to determine appropriate responses to degradation, and to establish responses that are particular to the places in which they occur. Technological restoration disconnects, commodifies, and emphasizes generality and interchangeability, whereas focal restoration grounds us in place and community.

Opportunities to cultivate ecological citizenship and to engage in grounding, focal practices are thus two important reasons to take seriously the participatory dimensions of ecological restoration. William Jordan III, longtime editor of the journal *Ecological Restoration*, argues that restoration can go even further in helping humans negotiate relations with the natural world. Jordan (2003) suggests that restoration is an activity that can help us grapple with the conflicted and ambiguous relationship between humans and nature, and particularly with the fact that as humans, we are both *part of* and *apart from* nature. We neither fit fully within nature, nor fully outside it: although we are animals and in a sense no less natural than any other creature on earth, we nevertheless experience nature as something other than or different from us. The tension in our relationship with nature is further accentuated by our ambivalence regarding our dependence on the natural world. Our dependence on plants and animals for sustenance accentuates our limits: we literally cannot survive without other organisms to eat. This dependency, suggests Jordan, is a source of shame, which he describes as

> a sense of existential unworthiness, the painful emotion a person naturally feels on encountering any kind of shortcoming or limitation, beginning with the infant's discovery that he or she is not omnipotent but is instead one of many others and dependent on those others for every kind of pleasure and satisfaction, and even for life itself. This shame is inseparable from any experience of relationship for the simple reason that any relationship forces on us an awareness of difference, and therefore of limitation. (Jordan 2003, pp. 46–47)

The shame we encounter in relation to nature is not something that can be resolved or overcome, argues Jordan, and we suppress it at our peril. The attempt to deny, rather than acknowledge, our limits leads us to seek liberation from nature, ignoring the fundamental reality of our embeddedness and reliance on it. However, without productive ways to acknowledge our limits, we can be paralyzed by a sense of unworthiness, attempting to save nature by standing apart from it. But as William Cronon (1995) has pointed out, an ethic of preservation that glorifies the pristine and treats all other landscapes as tainted may leave no place for humans to constructively interact with the natural world. In this sense, it is not much of an ethic at all: we attempt to save ourselves from moral errors by withdrawing from interaction. But this is as ludicrous in the case of nature as it would be in the human case: an ethical life cannot be achieved by avoidance. Ethics is fundamentally about relations. What's more, we cannot *avoid* nature, we are embedded in it.

Restoration offers a way out of the dilemma created by denying or failing to deal productively with our limits as human beings. In Jordan's view, we can think of restoration not only in terms of the changes it effects on the ground, but as a kind of active, performative enterprise in which we negotiate our relations with nature. As such, restoration can dramatize our limits and highlight the ambiguities in human–nature relations.

As noted above, restoration itself is fraught with moral ambiguity. As both Katz and Higgs suggest, restoration can become another source of human domination. Processes of restoration almost always involve some degree of destruction: cutting trees, pulling weeds, or even killing off particular animals to make way for others. The performative aspect of restoration can highlight ambiguities and help us recognize them. For example, restoration and art can be integrated in ways that bring out the complexity of the restoration process. During prairie restoration work at University of Wisconsin Arboretum in the early 1990s, artist Barbara Westfall used art to draw attention to the removal of aspen trees from the site. She highlighted the girdling of the trees by stripping additional bark, and used paint and oil to draw attention to this grove of dying aspens, which were being removed to maintain the restored native prairie. Projects such as Westfall's can provide forms of engagement in restoration that encourage us to confront the challenges of our relationship to nature. Activities such as this celebrate the possibility of restoration as a way of renewing the land, while at the same time revealing "the discrimination, destruction and killing involved in restoration . . . and within the larger cycle of our taking from and giving back to nature" (Jordan 2003, p. 188).

For further thought

1. Are technological restoration and focal restoration both important forms of restoration? Is one preferable to the other?
2. Of the rationales for public engagement in restoration presented here, which is most compelling, and why?

Conclusion: Narrative, continuity, and the future of restoration

Ecological restoration can serve as a microcosm for exploring human relationships with nature more broadly. In part, this is because there are many forms of restoration, just as there are many aspects of our relationship with the natural world. Restoration occurs on scales that range from small patches the size of someone's backyard to watersheds encompassing thousands of square miles, as in the restoration of the Florida Everglades. Restoration can focus on ecosystems as they existed prior to significant human disturbance, or prior to a *particular* human disturbance, but they can also incorporate human activities and ongoing use. Restoration can make visible multiple layers of a place's socioecological history, or it can foreground one of these layers and background the rest. Restoration can emphasize the return of the "original" species to a site, or it can use species that are importantly analogous to these species, as in the projects in the Netherlands that introduce Heck cattle and Konik horses as functional equivalents of extinct, historic grazers. Restoration can be highly "technological" and dependent on professional expertise or it can be a grassroots, volunteer effort—or it can incorporate elements of both.

The challenges and opportunities associated with ecological restoration show that the human relationship with nature is not a simple problem to be solved, but rather requires ongoing effort and reflection. Our solutions to the challenge of living well in relation to the natural world will always be tentative and provisional, subject to reevaluation as we learn more and as contexts change. In recent decades, land managers have recognized the importance of adapting to new circumstances and integrating new information, embracing an approach known as *adaptive management*, which is based on a model of thoughtful experimentation, evaluation, and readjustment (Figure 7.1). In his book, *Ignorance and Surprise*, Matthias Gross (2010) proposes just this approach to ecological restoration. It is worth considering that this kind of tentative, experimental, and adaptive approach may be relevant not only in managing ecological systems, but in thinking through our ethical relationship with the natural world more generally. When applied thoughtfully, an adaptive approach can embody humility and recognition of the limits of our knowledge and understanding. An adaptive approach need not shy away from engagement and intervention, but it does critically examine those forms of engagement and intervention. If we learn, for example, that our approaches to ecological restoration fail to take adequate account of environmental justice or local ecological knowledge, then we should take steps to remedy these shortcomings. An integrative, adaptive approach is one that not only takes account of new data regarding ecological conditions, but also engages with the diverse constituencies involved in and affected by ecological restoration.

Looking forward, ecological restoration will continue to grapple with the challenges associated with selecting reference conditions and defining restoration goals. These choices involve values: nature alone cannot tell us what to do. This point has been underscored by the proposal to focus on

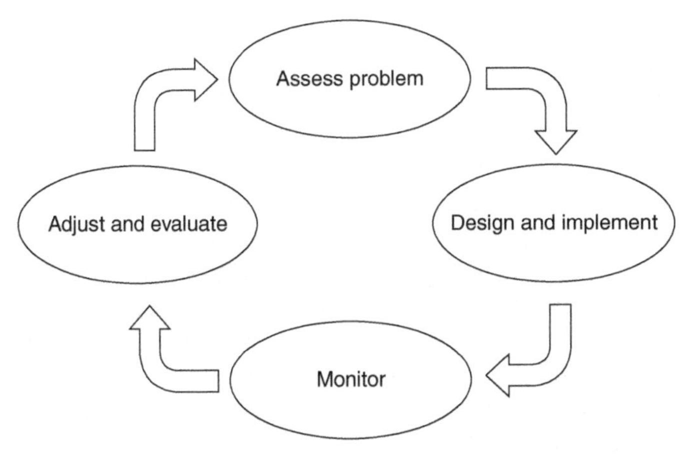

FIGURE 7.1 *Adaptive management. The process of adaptive management involves the development of management plans or policies, followed by implementation, monitoring, reassessment, and readjustment. Adaptive management is a cyclical process of learning and adaptation. Adapted from ch. 1 in Williams, Byron K., Robert C. Szaro, and Carl D. Shapiro. 2009.* Adaptive management: The U.S. Department of the Interior technical guide. *Adaptive Management Working Group, U.S. Department of the Interior, Washington, DC. http://www.doi.gov/initiatives/AdaptiveManagement/TechGuide/Chapter1.pdf*

Pleistocene re-wilding in North America, suggesting that we restore camels, cheetahs, elephants, and other living relatives of the extinct large mammalian species that roamed the continent 13,000 years ago (Donlan et al. 2006). While Pleistocene re-wilding strikes many as an outlandish idea, it does provoke us to consider explicitly why we choose one historic reference condition over another—and whether historic reference conditions are relevant at all. The role and place of history in restoration will, undoubtedly, continue to generate significant debate.

In this regard, we might consider an approach that incorporates history in a way that is dynamic, taking into account the interactions between humans and their environments over time. Think about the place where you grew up. How would you tell the story of that place? And how might that story shape your thinking about that place's future? A number of scholars have begun to explore the possibility that *narratives* may be useful in thinking through the possibilities for restoration at a particular site, suggesting that narratives bring out the ways in which people value places and how various values are interconnected. Rather than *enumerating* values, narratives help us contextualize them. Narratives also have a temporal dimension—they describe the story of a place over time—and this gives them a directionality that may help us think through future possibilities, and how these possibilities build on the past. Philosophers John O'Neill, Alan Holland, and Andrew Light

(2008) suggest that the narrative of a place—which incorporates ecological as well as social and cultural dimensions, and their interaction—can tell us what *ought* to come next, and how we *ought* to restore. There are challenges with this approach, as we can tell multiple, conflicting stories about a particular location, and environmental conflicts often center around conflicting narratives (McShane 2012; Hourdequin 2013). Nevertheless, in places with important socioecological meanings, narratives may provide insight into how various restoration options would embody, or fail to embody, particular values and meanings. Narratives bring to restoration testimonial knowledge—or "knowledge of direct conviction" based on experience—which can serve as a counterbalance to paradigmatic knowledge, which relies on generalized applicability and is characteristic of the knowledge embodied in scientific theories (Higgs 2003, p. 200).

In a case study of wildland fire management near San Diego, California, Bruce Goldstein and William Butler (2010) found promise in a narrative approach to understanding competing visions and collaboratively working to reconcile them. In this case, they explain,

> [C]ompeting narratives were markers of incommensurability of institutional order, ways of knowing, and professional identity. Yet narratives can also provide collaborators with insight into each other's perspectives and values, a way to grapple with complexity and uncertainty while expressing individual and collective identity. Developing collective narratives permits participants to reassemble familiar ideas, methods, and strategies, trying different combinations until a new story emerges that seems workable and mutually acceptable. (Goldstein and Butler 2010, pp. 6–7)

A narrative approach to ecological restoration, therefore, may be most useful in conjunction with a collaborative process in which competing narratives are identified and discussed, and new narratives explored. The narrative approach is not an algorithm for setting restoration goals, but given the discussions in this chapter, no algorithm can adequately account for the contextual and place-specific character of restoring ecological—and socioecological—systems.

Further reading

Choi, Y. D. (2007). "Restoration ecology to the future: A call for new paradigm." *Restoration Ecology,* 15(2), 351–353.

Donlan, C. J., Berger, J., Bock, C. E., Bock, J. H., Burney, D. A., Estes, J. A., et al. (2006). "Pleistocene rewilding: An optimistic agenda for twenty-first century conservation." *American Naturalist,* 168(5), 660–681.

Elliot, R. (1982). "Faking nature." *Inquiry,* 25, 1–93.

Higgs, E. (2003). *Nature by Design: People, Natural Process, and Ecological Restoration.* Cambridge, MA: MIT Press.
Hourdequin, M. (2013). "Restoration and history in a changing world: A case study in ethics for the Anthropocene." *Ethics and the Environment,* 18(2), 115–134.
Hourdequin, M. and Havlick, D. G. (2013). "Restoration and authenticity revisited." *Environmental Ethics,* 35(1), 79–93.
Jordan, W. R. III (2003). *The Sunflower Forest: Ecological Restoration and the New Communion with Nature.* Berkeley, CA: University of California Press.
Jordan, W. R. III and Lubick, G. M. (2011). *Making Nature Whole: A History of Ecological Restoration.* Washington, DC: Island Press.
Katz, E. (1997). "The big lie: Human restoration of nature," in *Nature as Subject: Human Obligation and Natural Community.* New York: Rowman & Littlefield, pp. 93–108.
Kolbert, E. (2012). "Recall of the wild: A quest to engineer a world before humans." *New Yorker Magazine,* December 24, pp. 50–60.
Light, A. (2006). "Ecological citizenship: The democratic promise of restoration," in R. H. Platt (ed.), *The Humane Metropolis: People and Nature in the 21st-Century City.* Amherst, MA: University of Massachusetts Press, pp. 169–181.
O'Neill, J., Holland, A., and Light, A. (2008). *Environmental Values.* New York: Routledge.

8

Engaging environmental concern, promoting change

Introduction

We have now almost completed our tour of environmental ethics. This last chapter brings together many of the themes discussed earlier as we consider the possibilities for ethical change and moral progress in relation to the natural world. We will begin with a discussion of environmental pragmatism, value pluralism, and environmental politics. This discussion also will raise a question: Do pragmatist approaches—which work from the existing values that people hold—retain the potential to genuinely transform our moral outlooks in relation to the natural world? In considering this question, we will explore the prospects for stretching our moral and environmental imaginations through literature and art, then link this discussion to the prospects for personal, cultural, and political change. At the end of the chapter, we will examine the challenges and opportunities for positive social change, as illustrated through a discussion of sense of place, the role of ritual in cultural change, and the possibilities for imagining and enacting an ecological culture.

The first part of the chapter introduces environmental pragmatism, building on two ideas from Chapter 7: adaptive management and narrative approaches to environmental ethics. Adaptive management is a process that relies on experimentation, monitoring, and readjustment as we learn about what works and what does not in acting to restore, manage, or interact with the natural world. Narrative approaches to environmental ethics emphasize the temporal dimensions of a place and the socioecological processes that have occurred there over time.

Both ideas—adaptive management and narrative—resonate with an approach to environmental ethics that attempts to marry theory with practice, or perhaps more precisely, attempts to develop theory *through* practice. This approach, called *environmental pragmatism*, arises from a critique of fundamental value theory in environmental ethics. Environmental pragmatists worry that environmental ethics tends to get bogged down in theoretical arguments over what intrinsic value is and what sorts of things possess it. These debates, they suggest, have the potential to result in interminable theorizing, at the expense of practice. I suggest later that there remains an important role for value theory in environmental ethics. Nevertheless, environmental pragmatism offers a valuable reminder that environmental ethics, at its core, poses a critical, practical question: How should we live in relation to the natural world? What's more, environmental pragmatism makes an important move by drawing environmental ethics into the political arena. This is key, because any practical environmental ethics must in some way be political.

Rather than focus on fundamental value theory, pragmatists take a problem-based and politically engaged approach to environmental ethics. Although many views fall under the broad umbrella of environmental pragmatism, Andrew Light and Eric Katz characterize it as follows:

> [E]nvironmental pragmatism is the open-ended inquiry into the specific real-life problems of humanity's relationship with the environment. (Light and Katz 1996, p. 2)

Returning to the ideas with which we began, what do adaptive management and narrative have to do with environmental pragmatism? An important part of the answer draws on an earlier point: environmental pragmatists attempt to develop theory *through* practice. Rather than begin with foundational value assumptions, environmental pragmatists begin *in media res*—in the middle of things—taking as starting points the diverse values that people actually hold.

We can see how this fits with a narrative approach to environmental ethics, which pays careful attention to particular places and the people who inhabit them and attempts to understand the complex socioecological interactions and values that evolve in those places over time. Like environmental pragmatism, narrative approaches look to existing values to better understand relationships between people and places, and the possibilities for the future, given current and historical contexts. As John O'Neill, Alan Holland, and Andrew Light (2008, p. 153) put it, "Ethical reflection needs to begin from the thick and rich ethical vocabulary we find in our everyday encounters with the environments which matter to us." Such an approach can help avoid a yawning gap between what theorists tell us we *should* value and what people *actually* value.

Of course, if our relationships with the natural world are flawed and imperfect—as much of environmental philosophy tells us they are—then we ought not rest easy with current values. Environmental pragmatists are aware of this, and suggest that although we should take existing values as starting points, these values can be transformed through thoughtful discussion in community and political contexts. This is where the idea of adaptive management comes in. In his book, *Sustainability: A Philosophy of Adaptive Ecosystem Management*, Bryan Norton (2005) argues that the idea of adaptive management from forestry and natural resource management can be expanded to encompass a broader social process of experimentation and readjustment. This broadened conception of adaptive management is inclusive and process-oriented. Not only "experts"—scientific, philosophical, economic, or otherwise—participate in deliberation about environmental policies and practices; Norton envisions a broadly engaged public.

In addition, Norton's view of adaptive management is place-based and context-sensitive, yet attuned to the way in which environmental problems involve multiple scales in space and time. Thus, adaptive management may begin from a local place, but at the same time consider the way that place interacts with *other* places. Along the temporal dimension, Norton's adaptive management approach emphasizes not only the short-term consequences of management decisions, but effects on future generations as well.

Adaptive management, in this view, involves ongoing social learning, and brings together discussions of value with discussions of "facts"—or more precisely, it emphasizes the way that values and facts are intertwined, and how both can be reassessed in light of experience. For example, one might think that urban growth is desirable for various reasons—it may help fuel overall economic growth, or keep housing prices low—but experience can help test the value of unbridled sprawl. As Norton (2005, p. 191) explains, "[M]any U.S. cities . . . have over the past few decades carried out an experiment in unlimited suburban growth. This experiment was animated by a value, the value individuals place on freedom and unlimited mobility." Yet people's values and desires in this case should be viewed as subject to confirmation—or disconfirmation—by experience. As Norton notes, time will tell whether people really value freedom and unlimited mobility as much as they thought they did, or more than other values that are at stake in urban and suburban development choices.

Pluralism, pragmatism, and politics

As our discussion thus far has shown, environmental pragmatism is a distinctive theoretical perspective that gives lived experience a central role and casts doubt on foundationalist theories of knowledge and value. (Foundationalist theories identify certain key beliefs or values as basic, or given, and justify other beliefs, values, or principles in relation to these fundamental foundations.) From a theoretical perspective, environmental pragmatism takes its cue from *philosophical pragmatism*, an earlier movement that developed in the late nineteenth and early twentieth centuries in the work of American philosophers such as Charles Pierce, William James, and John Dewey.

Pragmatists reject the foundationalist aims of establishing a firm and unchanging basis for epistemology and ethics. Instead, pragmatist epistemology holds that the truth of a belief should be judged in relation to experience. In the ethical realm, pragmatists begin with what people actually value in their lives. Because people value many things in many ways, pragmatists are typically pluralists, recognizing many legitimate values rather than any single overarching moral value or principle from which all others are derived. Pragmatists view ethical theories as provisional and subject to revision, understanding "ethics as a process of continual mediation of conflict in an ever changing world" (Parker 1996, p. 25). As we have seen, the aim of pragmatist ethical theories is to make sense of the diverse, interrelated values that people hold, and to engage in an ongoing, recursive process of evaluating and reevaluating those values.

Environmental pragmatism: Plurality and process

As noted earlier, a key aim of environmental pragmatism is to make environmental ethics more practical and politically engaged, providing a counterpoint to seemingly interminable debates about intrinsic value in environmental ethics. Pragmatists offer two key critiques of intrinsic value theories. First, they see these theories as problematic in their a priori attempts to pin down fundamental bases for value. Such aims fit poorly with the pragmatist conviction that values are plural and contextual, emerging from lived experience. Second, environmental pragmatists worry that a "theory first" approach will remain forever embroiled in theory, and never get to practice.

Environmental pragmatists argue that the search for a single, foundational set of values is misguided (see, e.g., Minteer 1998). Instead, we should consider the ways in which people actually value animals, plants, and natural world, and work from there. Rather than focus on establishing rock-solid value foundations, we should view values as a complex, interconnected web, in which values in one part of the web support those in other parts (Weston 1996). Pragmatists embrace *value pluralism*—the idea that there are multiple, legitimate values—and tend to reject hierarchical views in which a single value can justify all others. This is, in part, because pragmatists generally do not see values as "out there" in the world, awaiting discovery by human beings; instead, they see values as something *we* create, through the activity of valuing.

Rejection of foundationalism goes hand in hand with the environmental pragmatists' second critique. If there are no objective, timeless values that can ground ethics, there is no sense spending decades—or centuries—seeking them out. If environmental philosophers care about environmental issues in the real world, they should engage in the value discussions happening there. Environmental pragmatists thus argue that a practical, problem-oriented environmental philosophy is in order (Norton 1996). Rather than working out particular value relationships in a philosophical ivory tower, pragmatists embrace a political and process-centered approach:

> [P]ractical philosophy does not assume that useful theoretical principles will be developed and established independent of the policy process and then applied within that process. It works toward theoretical principles by struggling with real cases . . . Practice is prior to theory in the sense that principles are ultimately generated from practices, not vice versa. (Norton 1996, p. 108)

Environmental pragmatism thus lends itself to a context-sensitive examination of environmental values, and to public participation and democratic engagement in elaborating ethical principles. Pragmatists see knowledge and values as provisional and revisable in relation to particular contexts.

The promise and limits of pragmatic politics

With its practical and problem-oriented approach, environmental pragmatism calls for renewed civic engagement and robust dialogue surrounding environmental issues and human–nature relations. This is an appealing feature, inviting increased attention to deliberative democracy—public and political conversation in which participants are open to the ideas of others, and to transforming their own views through social learning. Under conditions of trust and mutual respect, such conversations can enable participants to move beyond the promotion of their narrow self-interests and take a broader perspective, together formulating a conception of the public good. From the perspective of environmental ethics, pragmatists such as Bryan Norton (1991) and Ben Minteer and Robert Manning (2000) have argued that the resultant "enlightened anthropocentrism" can generate policy outcomes that converge with those recommended by a nonanthropocentric point of view. This *convergence hypothesis* holds that in the practical realm, enlightened anthropocentrists may come to the same conclusions as nonanthropocentrists.

As Bryan Norton (2005, p. 273) puts it, "When the goal is cooperative management of shared problems, consensus on what to do is more important than consensus on why it is important to do it."

Although it is not clear that anthropocentrists, biocentrists, and ecocentrists always will agree on matters of policy, there is some evidence that a pragmatist approach has the potential to take environmental concerns seriously. A study of Vermonters' environmental attitudes, for example, found that citizens hold a diverse array of environmental values, with utilitarian conservation, stewardship, and strong environmental attitudes all well represented among the public (Minteer and Manning 2000). This study, at least, suggests that greater public involvement in environmental decision-making would not necessarily promote short-sighted, human-centered policies, but could express concern for future generations, the contribution of the natural world to the quality of life of humans, and our strong dependence on nature (Minteer and Manning 2000).

Environmental pragmatism also fits well with increasing reliance on collaborative processes involving multiple stakeholder groups in environmental decision-making. As we saw in Chapter 5, collaborative Food Policy Councils work for systemic improvements in local and regional food systems by involving citizens, public officials, farmers, and others involved in food production, transport, and sale. To take another example, many federal land management agencies in the United States, such as the U.S. Forest Service, have developed collaborative working groups, with significant success. Although these groups have their limitations, they often help diverse constituencies to understand one another and to work toward creative resolutions of land management controversies. Collaborative forest management groups, for example, have begun to develop approaches that protect watersheds and sensitive species while selectively harvesting trees.

Environmental pragmatism favors social change through the model of the "connected critic" introduced in Chapter 1. In this model, social critics seek resources for reform in existing values and traditions. Poet and essayist Wendell Berry, for example, draws on values central to the agrarian tradition in his arguments for reconnection with the land. As Berry writes in *The Unsettling of America*:

> I believe that the answers are to be found in our history . . . I am talking about the idea that as many as possible should share in the ownership of the land and thus be bound to it by economic interest, by the investment of love and work, by family loyalty, by memory and tradition . . . The old idea is still full of promise. It is potent with healing and with health . . . It proposes an economy of necessities rather than an economy based upon anxiety, fantasy, luxury, and idle wishing. It proposes the independent,

> free-standing citizenry that Jefferson thought to be the surest safeguard of democratic liberty. And perhaps most important of all, it proposes an agriculture based upon intensive work, local energies, care, and long-living communities . . . (Berry 1997, p. 14)

Here and elsewhere, Berry clearly appeals to American history in an effort to unearth and revitalize what he sees as long-standing cultural values. Berry engages in a form of social activism that suggests that Americans are abandoning their own traditions, and that they have allowed values such as convenience and efficiency to eclipse commitments to family, community, and the land.

Such approaches can be powerful: for those who embrace their cultural heritage, values central to that heritage can carry with them a sense of familiarity and stability. Yet questions remain. Can existing traditions provide the conceptual resources needed to address our current circumstances? Are such traditions sufficiently rich to guide the development of a robust environmental ethic? We have already seen that the discipline of environmental ethics emerged in direct response to the perceived *in*adequacy of Western moral, cultural and religious traditions, which tend to place human beings at the center and give little attention to the natural world. Thus it seems that at the very least, environmental ethics will need to call on values that have been backgrounded in the Western tradition, or extend and reimagine the values central to its core.

The appeal of environmental pragmatism is that it validates the values we already hold, viewing them as resources for the establishment of more thoughtful policies, attitudes, and actions in relation to the natural world. Yet, pragmatism risks conservatism in working from existing values, and one might worry that environmental pragmatists have been too quick to dismiss the significance of intrinsic value specifically and philosophical theory more generally (McShane 2007b, Samuelsson 2010). Theories themselves can play a role in political discussions, by offering a coherent way of valuing, developing a particular moral outlook, and systematizing and expressing certain inchoate moral intuitions. One vivid example of this comes from the animal rights movement: Peter Singer's book *Animal Liberation* has sold more than half a million copies, and part of its popularity no doubt derives from the way the book gives voice and structure to the intuitions of many animal rights activists that the amount of suffering involved in industrial animal agriculture and animal experimentation is morally wrong.

Philosopher Andrew Light (2002) suggests a possible solution to the apparent dilemma between pragmatism's potential conservatism, on the one hand, and the political disconnectedness of moral theory, on the other. He proposes what he calls *methodological environmental pragmatism*.

Though the term is a bit of a mouthful, the basic idea is this: there remains a place for nonanthropocentric environmental theory, but at the same time, environmental philosophers should offer arguments that can effectively motivate anthropocentrists in the political arena. In this approach, one might represent nonanthropocentrism in one context, and (enlightened) anthropocentrism in another. The worry, of course, lies in the potential schizophrenia of this strategy. How are such divergent forms of argument to be reconciled? In some cases, they may complement one another; in other instances, they may conflict.

An alternative approach might be to commit oneself to one's deepest and most genuine moral convictions, then employ creative strategies to stretch the public moral imagination to take seriously what initially might seem like radical views. Pragmatists raise legitimate concerns about the relationship between theory and practice, and between values and action. However, theories play a variety of roles in thought and understanding, and perhaps not all ethical theories, or even all environmental ethical theories, need to focus on questions of practice, action, and moral motivation. Nevertheless, if environmental philosophy is to play an active role in promoting change, then we need to take seriously questions of motivation and action, and seek creative ways to bring theory and practice into lively conversation with one another.

For further thought

1. Is the pragmatist critique of intrinsic value theory on target? Should environmental ethics give up on foundational value theory?
2. What are the core strengths and weaknesses of environmental pragmatism?

Enlivening the moral imagination

As suggested earlier, we seem to face a dilemma in thinking through the connection between theory and practice in environmental ethics. People best understand and are most easily motivated by values they already hold, and pragmatic approaches to ethics have the advantage of taking existing values seriously. However, many current value priorities—efficiency, unbridled freedom, and ever-increasing material wealth—align poorly with respect for the natural world and care for future generations. Thus, restricting ourselves to existing values may limit significant social change. On the other hand, biocentrism and ecocentrism suggest a strong reorientation of values, but because of their limited connection to current values and institutions, may fail to motivate action.

Social change is a complex process, and it seems fair to say that no one fully understands it. Yet most effective social movements and leaders of change draw both on existing values *and* call us to embrace new ones, or at least new ways of thinking about the old. Environmental ethics does something similar, asking us not only to embrace conventional values, but to extend our moral imaginations and even to *think the unthinkable*. In his essay, "Should trees have standing," Christopher Stone (2010) begins with a section called "the unthinkable," in which he notes that many ideas we now take for granted—that children are persons, for instance—were once beyond the pale. At one time, Stone tells us, a child was considered merely "an object, a thing" that could be owned. Similarly, Peter Singer (2002) notes that in the late eighteenth century, Mary Wollstonecraft's arguments for the rights of women were ridiculed as absurd, though in restrospect, they seem prescient. Although we are inclined to view ideas that challenge our standard ways of thinking and acting as "unthinkable," history shows that we are capable of change. Aldo Leopold stresses this point when he discusses ethics as a process of gradual evolution, taking into consideration ever wider circles of persons, and ultimately, ideally, the land.

One of the central purposes of philosophy is to invite reflection on the assumptions that we take for granted, and to subject those assumptions to critical scrutiny. We tend to grow comfortable and confident in our beliefs, until it seems simply *obvious* that the way we do things is *the* way to do things, and that there really is no other way. Reflection can reveal the limitations of these assumptions, and when we engage with unfamiliar cultures or read about other times and places, we find an incredibly diverse array of possibilities for living in the world. Philosophy can catalyze reflection and loosen the grip of our predispositions, but arguably, we need more than philosophy to catalyze a shift in values and healthier, more sustainable relationships with the natural world, and with one another. We need ways of connecting not only intellectually, but also affectively with other human beings and other living things, and we need modes of thought that enable us to reimagine the possibilities for living well in the world. Below we consider three ways to enliven our moral and environmental imaginations: through direct experience; literature; and art, design, and the aesthetics of everyday life.

Experience

Aldo Leopold (1949, p. 214) once said that "we can be ethical only in relation to something we can see, feel, understand, love, or otherwise have faith in," and he argued that standard approaches to conservation education may be insufficient to develop a land ethic; instead, a particular *kind* of education is needed (Leopold 1949, pp. 207–208). Leopold insisted that we need not only education of the head, but also education of the heart. Leopold's writings suggest that developing the land ethic requires a love for the natural world, for living things, and for the beauty and complexity of ecological communities. But how *does* one educate the heart?

The lives of great naturalists and conservationists suggest that experience is key. In *A Sand County Almanac*, Leopold chronicles a year at his family's sandy, degraded Wisconsin farm through a series of vignettes capturing his careful observation and engagement with the natural world. He is attentive even to the smallest flower, noting:

> He who hopes for spring with upturned eye never sees so small a thing as Draba. He who despairs of spring with downcast eye steps on it, unknowing. He who searches for spring with knees in the mud finds it, in abundance.
>
> . . . Draba plucks no heartstrings. Its perfume, if there is any, is lost in the gusty winds. Its color is plain white. Its leaves wear a sensible woolly coat. Nothing eats it; it is too small. No poets sing of it. Some botanist once gave it a Latin name, and then forgot it. Altogether it is of no importance—just a small creature that does its job quickly and well. (Leopold 1949, p. 26)

This passage illustrates Leopold's capacity to notice the unnoticeable, and to care about living things that others would easily miss or trample without a thought. Calling the plant a "creature" dignifies it, for this term is usually reserved for animals, generally considered "higher" organisms. Leopold's language has an implicit leveling effect: all are important, even those we often overlook.

In Leopold's writings, and also in those of scientist, naturalist and twentieth century writer Rachel Carson, we see a profound affective connection with other living things. Carson suggests that the capacity to connect with the world and to see it as wondrous and beautiful is something we all possess as children, though it often grows duller over time:

> A child's world is fresh and new and beautiful, full or wonder and excitement. It is our misfortune that for most of us that clear-eyed vision, that true instinct for what is beautiful and awe-inspiring, is dimmed and even lost before we reach adulthood. If I had influence with the good fairy who is supposed to preside over the christening of all children, I should ask that her gift to each child in the world be a sense of wonder so indestructible that it would last throughout life, as an unfailing antidote against the boredom

> and disenchantment of later years . . . the alienation from the sources of our strength. (Carson 1965)

The alienation that Carson talks about is facilitated in part by our physical and material disconnection from the natural world. As Leopold (1949, pp. 223–224) explains, in modern life we are separated from the world by "many middlemen" and "innumerable physical gadgets." This is no doubt truer now than when written more than six decades ago. Although reconnection with the natural world is undoubtedly a complex and multifaceted challenge, direct experience is a good starting point, and may indeed be a crucial prerequisite for an environmental ethic with genuine motivational power. The sheer richness of direct experience—engaging diverse senses through sound, sight, touch, smell, and even the taste of the air or the snow on one's tongue—can leave an indelible mark. Experience, argues contemporary writer and environmental educator Mitchell Thomashow, engenders wonder, and "[a] state of wonder is the basis for an ethic of care" (Thomashow 2002, p. 57).

The words of Leopold, Carson, Muir, Thoreau, or other iconic nature writers can inspire us, but our connections to such writings seem to depend on some prior knowledge and experience which literature can reactivate and deepen. Thus, a critical step to an enlivened moral and environmental imagination may be direct experience and attentive, receptive encounters with both human and nonhuman worlds.

Literature

In writing about experience, I have already drawn on literature to assist. The lives and writings of Rachel Carson and Aldo Leopold exhort us—both directly, and by the example of their own lives—to get out into the world, and to experience it, to notice what is there. Through attentiveness and a childlike openness to the world, we can come to care more deeply for what surrounds us, and what we so often overlook.

Literature can also invite us to imagine the world from the perspective of other living things, and to turn ourselves outward, as in these opening lines of a poem by Mary Oliver (1997, p. 61):

> Have you ever tried to enter the long black branches
>   of other lives—
> tried to imagine what the crisp fringes, full of honey,
>   hanging
> from the branches of the young locust trees, in early summer,
>   feel like?

Oliver later comments (to those who haven't):

> No wonder we hear, in your mournful voice, the complaint
> that something is missing from your life!

Poetry, essays, short stories, and science fiction also can draw us into worlds—real and imagined—that we might never directly experience. They enrich our sense of possibility and, in philosophical terms, help us "think counterfactually," imagining ways in which things could be otherwise. Literature can present utopian and dystopian worlds, either beckoning or warning us away. Considering these possibilities helps us work through different horizons of time and place. We can travel without going anywhere, and return with a different perspective on our own world and role in it.

Literary works often raise important philosophical and ethical questions, and many philosophical writers make effective use of literary techniques. "Thought experiments" in philosophy, for example, ask us to consider possible situations and choices different from our own, and reflect on our intuitions about these cases as a way of teasing out ethical questions and testing various ethical principles. Other philosophical work goes further. For example, Tim Mulgan's book, *Ethics for a Broken World* (2011), employs a framework that takes us out of our own time and provokes reflection on the moral and political philosophies of the present, "affluent age," from the perspective of those living in a dystopic future world.

Mulgan's book is presented as a series of lectures by a professor in a world that has been deeply altered by catastrophic global climate change. It thus asks us to consider the adequacy of our own moral and political philosophies from the perspective of future generations, and how the values and guidance they offer may be inadequate to the challenges we currently face. *Ethics for a Broken World* is powerful because it transports us to a different time, and the effect of that move is to broaden and enrich our understanding significantly more than standard arguments about obligations to future generations generally do. As one reviewer (Nathanson 2012) writes, "As a reader, I was left with a much more vivid sense of the need both to confront climate change as a problem and to think more seriously about the lives of the future people who will follow us." Environmental philosophy, therefore, may benefit from engagement with literature and literary techniques in the effort to imagine and make vivid new possibilities for our relationships with the natural world.

Art and aesthetic engagement

Like literature, other art forms can challenge our standard assumptions and may literally change our ways of seeing the world. Although we may tend to think of art as paintings hanging on a wall or sculptures in a museum, there are many forms of art and many forms of aesthetic experience. Appreciating the beauty of the natural world can be a critical starting point, tying back to the experiential dimensions discussed above. Leopold himself insists on aesthetic appreciation as an important basis for an ethical relationship with the natural world. In "Marshland elegy," he writes:

> Our ability to perceive quality in nature begins, as in art, with the pretty. It expands through successive stages of the beautiful to values as yet uncaptured by language. (Leopold 1949, p. 96)

It is not uncommon that aesthetic experience can be transformative, whether it be an encounter with the grandeur of an old-growth redwood grove, a cardinal so bright it seems to glow, or the aesthetic shock of a clearcut swath or a gaping, open pit mine. What is interesting about Leopold's view is that he very clearly ties aesthetic appreciation to knowledge of ecology and natural history (Callicott 1983). Such knowledge can deepen and enrich perception of natural places and the organisms that inhabit them. A marsh—which we might first be tempted to see as dreary and uninviting—can be transformed by an understanding of its evolutionary history or the ecological relationships that exist there. Leopold's writing clearly draws out the way in which a crane marsh captures the depth of time:

> A sense of time lies thick and heavy on such a place. Yet since the ice age it has awakened each spring to the clangor of cranes. The peat layers that comprise the bog are laid down in the basin of an ancient lake. The cranes stand, as it were, upon the sodden pages of their own history. (Leopold 1949, p. 96)

In the field of environmental aesthetics, perspectives such as Leopold's are known as *cognitivist* views. According to cognitivists, it is not our untutored emotional reactions to landscapes or natural places that lead us to fully grasp their aesthetic qualities; instead, knowledge and understanding of those landscapes and places play a critical role. What's more, the cognitive and affective dimensions of aesthetic experience interact. Knowledge can enhance one's emotional response to a landscape in various ways: by changing our interpretation of what we see, hear, or smell, or by changing—quite literally—what we perceive. One of the critical, but perhaps underappreciated, elements of the study of science and natural history is the training of perception. Go into the woods with a knowledgeable birder: where the uneducated ear hears a mass of birds "tweeting," the birder recognizes a panoply of distinct individuals and a diverse array of species. Only the trained ear can fully aesthetically appreciate the diversity of avian life present in that woods—the rest of us simply fail to notice what's there.

Those who study environmental aesthetics are often critical of the way in which aesthetic theory centers around art and historically has given little attention to the aesthetics of the natural world. Due to the dominance of art-centered aesthetics, even the appreciation of natural beauty is often filtered and at times distorted by aesthetic standards for art. For example, the tradition of landscape painting and the associated ideal of the picturesque "proposes that we should aesthetically experience nature as we appreciate landscape paintings" (Carlson 2012). However, this leads us to perceive the natural world as primarily two-dimensional (Carlson 2012), and to emphasize visual aspects of nature experience over aspects perceived through other senses such as touch, hearing, or smell (Callicott 1983, p. 346).

Aesthetic theory based on art—as well as art itself—can limit our understanding of and relationship with nature. However, art can also play a variety of roles in stimulating environmental and ethical imagination. For example, it can accentuate and make vivid various natural processes, including natural processes managed or manipulated by human beings. As noted in Chapter 7, artist Barbara Westfall's work provides a powerful example. In "Daylighting the woods," Westfall stripped the bark and darkened the trunks of aspen trees in Wisconsin that had been girdled to make way for the restoration of native prairie (see http://www.barbarawestfall.com/daylighting.php). Westfall's work does not reject prairie restoration, but it does reveal the complexity and moral ambiguity of restoration processes, showing how death makes way for new life.

Art can also provide forums in which to envision, model, or try out new relationships between humans and nature. Lillian Ball's WATERWASH® projects offer creative and aesthetic modes of stormwater remediation, integrating art and ecological design. In one of these projects (see http://www.lillianball.com/waterwash/doc/Abstract_WATERWASHtm_ABC.pdf), Ball created a wetland along the edge of the Bronx river in New York, using native plants to slow and reduce stormwater runoff through greater infiltration and water retention by vegetation. As in her other WATERWASH® projects, Ball designed this art/mitigation project to include walking paths constructed of permeable recycled glass. The site also incorporated interpretive signs to educate the public about the purpose and function of the project. Ball's artwork both serves a direct, practical purpose and offers an invitation to rethink urban design and standard approaches to stormwater management.

Similarly, artist Patrick Dougherty's willow sculptures may help us rethink our built environments and their relationship to the natural world. Dougherty constructs intriguing large-scale sculptures out of willow or maple saplings. These projects are often installed on college campuses, at museums, or in other public spaces, and feature tunnels or rooms through which people can walk or children can play hide and seek. The sculptures gradually degrade over time, highlighting gradual processes of change and decomposition in nature. Whether intended or not, Dougherty's work helps us envision other possible modes of construction, such as houses grown from trees, as architect and urban designer Mitchell Joachim suggests (see http://www.archinode.com/fab-tree-hab.html).

Art can thus broaden our imagination and help us envision new possibilities. It can challenge dominant paradigms and reveal their limitations. Finally, art and aesthetic engagement can contribute to the development of an ecological culture. Such a culture requires forms of aesthetic appreciation that go beyond art, encompassing the natural world as well as the spaces we inhabit every day. Considering the aesthetic dimensions of everyday life and their relationship to the natural world encourages us to transform cultural spaces into *eco*cultural spaces, and to consider more thoughtfully the aesthetics of everyday life (see Saito 2007).

For further thought

1. What experiences, books, works of art, or aspects of the built environment have influenced your relationship to the natural world?
2. Describe one way in which creative use of the arts can spur changes in values, institutions, or the built environment to improve human relationships to the natural world.

Integrating the personal and the political: Toward an ecological culture

Much of ethical theory has focused on individual action, character, and intention, and it has also focused exclusively on human relations, paying little attention to the wider world in which we are embedded. The development of an ecological culture requires changing both of these orientations. We need ethical theories that attend to the social and collective dimensions of human thought and action, and that consider socioecological relations in all their depth and complexity. This means we need to look not only at human–nature relationships, but the way in which human social relationships themselves are shaped by our interactions with other living things and with natural and built environments. From this perspective, *any* comprehensive ethical theory should have something to say about nonhuman animals, plants, ecosystems, and the natural world more generally. Yet few general ethical theories—even contemporary ones—address the natural world. There clearly remains significant theoretical work to be done.

However, environmental ethics is not only about theory; it is also about practice. Philosophers tend to focus on *conceptual* frameworks that can guide thought and action; however, an ecological culture also requires educational, architectural, sociocultural, political, and economic frameworks that support ethical relations with the natural world. As we have seen, making ethics practical requires thinking beyond concepts to institutions, and to the ways in which art, literature, and experience can broaden our moral imaginations and make possible the conceptualization of new forms of life. Environmental ethics, in its broadest sense, involves many disciplines, professions, and perspectives. Development of a lived environmental ethic is clearly not a task for philosophers alone.

One of the biggest challenges in developing and realizing a robust environmental ethic lies in the absence of strong historical foundations (particularly in the Western tradition) and the dominance of contemporary values and practices inconsistent with such an ethic. Even from a narrowly anthropocentric perspective, our environmental problems are significant, and human self-preservation alone demands a reorientation of priorities. When we examine closely our current ways of life and reflect carefully on the ethical dimensions of our relationship to the natural world, we typically find a yawning gap between theory and practice, or ideal and reality. How can this gap be bridged?

From theory to practice: Obstacles and openings

To answer this question, we need to examine, acknowledge, and work to overcome the obstacles to an ecological culture. An ecological culture is one whose values and practices enable thoughtful, respectful, attentive, and sustainable relationships with other living things, and with the natural world as a whole. So, what are the obstacles? Environmental philosophers and others have identified a number of them. Philosophers and other scholars often frame the obstacles in the language of virtue and vice: we are limited, for example, by apathy, greed, gluttony, arrogance, denial, and procrastination (Cafaro 2005; Norgaard 2011; Andreou 2007).

There is no unified prescription for each of these ills, though various suggestions have been made. Dale Jamieson (2007) suggests that we cultivate the "green virtues." Others concur, yet emphasize that the virtues need to be social or collective virtues, because environmental problems require collective, political, and public engagement and action (Treanor 2010). Carol Booth (2009) takes a slightly different tact, arguing that environmental ethics needs to take a "motivational turn," focusing on the psychological bases for human action and bringing scientific insights to bear. Sociologist Kari Norgaard brings together the motivational questions with social ones, suggesting that motivation is socially and emotionally grounded, and shaped at multiple levels:

> Cognition, awareness, and denial emerge from a process that spans micro-, meso-, and macrolevels of social structure. At the microlevel, we see the role of individual emotions and interactions. At the mesolevel of culture, norms of behavior from conversation, feelings, and attention shape and reinscribe what is considered "normal" to think about, talk about, and feel. And these local cultural norms are in turn connected to larger political economic relations. (Norgaard 2011, p. 210)

Denial is a genuine barrier to motivation and action, but denial is not just an individual-level problem. Instead, argues Norgaard, denial is socially constructed and reinforced, and it is complex. We may recognize the need to change at one level, yet deny it at another. In light of these considerations, it is clear that developing a robust environmental ethic—and an ecological culture that embodies it—depends on processes at many different levels. What's more, we should keep in mind that ethical and cultural changes interact, and neither is categorically prior to the other.

In seeking to construct and live an environmental ethic, to repair our relationships to the natural world, understanding the obstacles is key. But how can we best navigate around them? In weaving our way through the thicket, we may do best to seek the openings rather than focus too closely on the barriers. This may help us avoid being overwhelmed by our shortcomings, or by the enormity of the environmental challenges we face. I have already tried to suggest a few of these openings: through literature, art, direct experience, and appreciation of the environments we inhabit, we can begin to reimagine our relationship with the natural world. Other openings are cultural, social, and political.

From a philosophical point of view, I have suggested throughout the book that a relational approach to ethics may offer another critical opening. A relational approach is appealing for a number of reasons. First, it does not focus on universal rules. It is doubtful that philosophers will discover a comprehensive set of rules that can guide our actions and relations to the natural world (or to one another, for that matter). The world is complex, and ethics must be sensitive to complexity and context. A relational approach depends on our ability to evaluate and judge our relationships with one another and with the natural world, but it is context sensitive, and does not emphasize rules. Second, a relational approach emphasizes a fundamental aspect of our humanity—that we are social animals—along with the fundamental ecological insight of interdependence and interconnectedness. Finally, a relational perspective builds on ancient roots, taking advantage of existing cultural resources in the mode of the "connected critic," which was introduced in Chapter 1. Both Aristotle and Confucius, for example, saw human beings as fundamentally social beings and suggested that humans realize their full human potential through relationships. Whether or not one subscribes to a strong, relational conception of the self, it seems clear that our ethical selves are importantly defined by the quality and character of our relationships to others. Might these "others" not include nonhuman others? Even more broadly, might the relationships critical to our moral lives not encompass our connections to the natural world?

Thinking along these lines can help us see that an individual's relationship to the natural environment is not merely a dyadic person–world relation. In fact, few, if any, human relationships are narrowly dyadic. Even relationships between life partners are bound up in myriad other relations: parents-in-law, children, mutual friends, neighbors, and so on. Similarly, my relationship to the natural world is neither abstract (it is not a relation to the "earth" or to the world at large, but rather to specific aspects and elements of that world), nor dyadic. Instead, my relationship*s* (now plural) to the natural world are concrete, place-based (though not necessarily limited to a *single* place), and mediated in significant and complex ways by my social relations, the institutions and infrastructure that surround me, and the buildings in which I live and work. Environmental ethics may be "personal," but not *only* personal.

A relational ethics offers a theoretical lens for considering the social and ecological dimensions of our moral lives in a single framework. It thus has the capacity to acknowledge the ways in which the social and the ecological are intertwined. This is important, because a lived environmental ethic is built, in part, through the development of an ecological culture, which is a distinctly social enterprise. In this way, relational approaches to ethics have the capacity to take seriously the ecofeminist point that social and environmental problems are linked: addressing the domination of nature requires an examination of domination as an aspect of culture more generally.

Bioregionalism, cosmopolitanism, and cosmopolitan bioregionalism

As I have suggested already, a relational ethical perspective has implications for practice, and for the construction of an ecological culture. For example, we can see through this lens why *bioregionalism*—which emphasizes connectedness to place, rootedness in place, and communities that are attentive to place—resonates with an outlook, such as that of deep ecology, that stresses self-in-relation. Rich, meaningful, and sensitive relationships require receptivity and time to cultivate. The history of European colonialism, in contrast, is primarily a tale of imposition, of remaking places and peoples according to imported cultural and aesthetic standards. This pattern is still with us, as those of us in the American West seek green lawns appropriate to lands much wetter than those in this arid region. Bioregionalism, championed by writers such as Gary Snyder, Peter Berg, and Wes Jackson, counters this tendency to impose upon the land rather than engage in conversation with it. Snyder (1990) recommends that we *reinhabit* the land, coming to know and care for the places we live; Jackson (1994) suggests "becoming native to this place," which requires, unsurprisingly, understanding in detail its character and particularities. Bioregionalism also recommends that we adjust our political boundaries to reflect natural boundaries. Watersheds, for example, might make more natural political units than the squares and rectangles whose boundaries are arbitrary from an ecological perspective. Bioregionalism further champions local economies, local food, and local identities. Consonant with a relational approach, bioregionalism champions deep, sustainable, respectful, and supportive relationships to the local communities and environments in which we live.

Yet, as appealing as the bioregional vision may be, we live in a globalized world, full of cross-continental interchange through travel and commerce. Contemporary cultures blend many diverse elements. Rather than hunker down, in place, perhaps we should celebrate our cosmopolitanism. Local relationships have the potential to be rich and deep, but they may also be narrow. Perhaps we should look outward in seeking an ecological culture, weaving an ever-broader network of relations, across the globe and beyond. Anthony Weston (2012) proposes something along these lines, in his vision of sustainable interplanetary exploration and "cosmic environmentalism" as elements of a newer, greener culture.

Must we choose? Local or interplanetary? Perhaps, we need both rootedness *and* interchange. Good relationships depend on some degree of rootedness and commitment, yet broader interactions can provide perspective: this is the benefit of contemporary cosmopolitanism. Mitchell Thomashow (1999) recommends a "cosmopolitan bioregionalism." As he explains:

> Global economy requires that bioregionalists explore both the immediate landscape (place) and those larger systems that exist beyond the horizon (space). The local landscape can no longer be understood without reference to the larger patterns of ecosystems, economies and bureaucracies. (Thomashow 1999, p. 126)

Places, identities, and relationships in this globalized world have transregional dimensions (Thomashow 1999, p. 129). Rather than lament this, we might celebrate it, without losing our grounding in place. We will have to be careful here, as there is a temptation to try to "have it all," when in fact we cannot.

We cannot have unlimited freedom to jet around the globe on holiday without at the same time damaging the planet. We cannot easily invest in our local communities and make a difference there if we are moving every 3 years.

Nevertheless, there is something to the idea of being locally grounded yet globally engaged. We are already embedded in a global web of connections and implications: economies, political processes, and cultures across the world are engaged in significant and ongoing interchange. If these exchanges and interactions are to reflect ethical values, we need greater awareness, understanding, and empathy for those whom our actions affect. A relational ethical perspective asks us to examine the quality and character of the relations most central to our lives—such as those with friends, family, and our local environment—but also to consider *all* of the relations in which we are embedded, as best we can, and to promote the development of institutions that enable ethical relationships at multiple scales.

Relational ethics and civic virtue

In addition to calling our attention to questions of place and bioregionalism in a globalized world, a relational perspective fits well with certain forms of virtue ethics. In particular, the relational approach resonates with views that emphasize the social, political, and cultural aspects of virtue. Although virtue ethics tends to focus on the character traits of individuals, virtues are habits of thought and action that shape and constitute our relations with other persons, other living things, and with nonhuman nature. On a spectrum from personal to political and social, virtues run the gamut. Although all virtues are arguably social in some way, some are very explicitly so. Whereas temperance rests closer to the personal end of the spectrum, virtues such as neighborliness and political engagement fall at the public and political end (Treanor 2010). Philosopher Brian Treanor (2010) argues that because environmental problems require political action, personal virtues alone are inadequate for the task that lies before us. Along the same lines, William Throop (2010) suggests that we need to cultivate "social sustainability virtues," virtues that enable us to engage in effective collaborative problem-solving. Certain virtues we typically celebrate—self-reliance, independence, and tolerance—may be too individualistic to address large-scale social and environmental problems.

On this view, we need to cultivate not only individual virtues, but *civic* virtues and the institutions that support them. Robert Bellah and his colleagues (Bellah et al. 2008), for example, argue that local political institutions can be critical in mediating the relationships between individuals and larger political processes. In the environmental arena, there are examples of successes. State and local governments, for example, have set their own greenhouse gas emissions targets, and in some cases, state and regional initiatives have led to important policy changes at the national level. For example, the state of Massachusetts, along with eleven other partner states, sued the United States Environmental Protection Agency (EPA) for failing to regulate carbon dioxide emissions under the Clean Air Act. The case went to the U.S. Supreme Court, and Massachusetts won, forcing the EPA to develop a strategy to regulate greenhouse gases. What enabled this victory? In this case, civically and environmentally concerned individuals worked together through state and regional institutions with the capacity to exercise political power.

From a cultural perspective, we need to think about how to cultivate environmental and civic virtues that integrate the local and the cosmopolitan. Aristotle offered important insights on the cultivation of virtue that remain significant today. He argued that *upbringing* and *habit* are key. Families and schools both play critical supporting roles in the development of virtues by encouraging and reinforcing ethical habits of mind and patterns of action. Relatedly, moral models and exemplars can inspire and guide us. However, cultivation of the virtues may benefit from additional scaffolding. In this regard, resources from classical Chinese philosophy may be helpful—even though these early thinkers themselves had little direct concern for building an ecological culture.

Ritual, perspective taking, and relationships

In Confucian philosophy, we find a deep emphasis on the sociality of virtue. Confucius suggests that the development of virtue begins in core familial relationships such as those between parents and children: these are the roots of moral personhood. The early Confucian philosopher Mencius argued that all humans possess the "sprouts" of virtue, which if nurtured correctly (yet not coaxed along too much), can grow to maturity. From a Confucian perspective, virtues learned in the family setting then extend to the broader community, and even to political life. Although Confucian philosophy theorizes more fully the ethical importance of the family, these general insights exist in various forms within the Western tradition. However, there is an important aspect of the Confucian tradition that Western moral philosophy has almost entirely overlooked: the significance of ritual, or in Confucian terminology, the *li*.

It may seem strange to turn to ritual in developing an ecological culture—for ritual in some ways seems quaint and old fashioned in our fast-paced, rapidly changing modern world. However, early Confucians argued that rituals are what give structure and expression to many of our most important values—and despite contemporary challenges to ritual and tradition, this remains true today. Respect, for example, is expressed through rituals such as bowing or shaking hands or polite forms of address. Gratitude is expressed through the ritual of the thank-you note. Rituals "link the participants to a general social and cultural order that [exceeds] the present time and space" (Larsen and Tufte 2003, p. 90). This characterization of ritual may lead us to think that ritual confines us to the status quo, reinforcing traditional ways of thinking and acting. Ritual *can* reinforce existing values and commitments, but it also has a creative dimension: rituals can provide opportunities for reflection, for negotiation of complex tensions, and for reinvention and redefinition of social structures (Larsen and Tufte 2003).

In the environmental arena, William Jordan III and colleagues (2012) suggest that ritual, which they see as a form of art, "provides a way to negotiate relationships and shape human subjectivity in a mindful, participatory, deliberative way." They go on to argue that ritual, along with other art forms, may be a key to enabling the kinds of "radical changes in ideas, values, and beliefs" required to reform human relations to the natural world. Ritual further allows us to confront the very real ambiguities of our relationship with nature: we are, inevitably, implicated in damage and destruction of life—even if only to eat. Ritual provides a setting in which we can acknowledge and grapple with this reality, while at the same time seeking to minimize the harm we cause.

Ritual can be conservative or radical, somber or celebratory. Arguably, we need all these elements as we renegotiate our relationship with the natural world and cultivate the virtues needed to live well in it. The early Daoist philosopher Zhuangzi—who mocked the seriousness and attachment to convention that seemed to characterize traditional Confucian rituals—might find something to agree with here, particularly on the celebratory side. Zhuangzi represents the think-outside-of-the-box counterpart to the Confucians' emphasis on order and tradition. His writings challenge us to shift our point of view as we move at a dizzying pace from the perspective of a small minnow to that of a giant bird, then back to that of a tiny bug. This perspective shifting suggests that we ought not take our own perspective too seriously, nor cling too steadfastly to tradition. Thus, the Confucian perspective on ritual order is nicely complemented by Zhuangzi's sense of spontaneity, creativity, and social critique.

Zhuangzi's multiperspectivalism also challenges our conventional assumptions about value, usefulness, and uselessness. Recall how Zhuangzi (2001, pp. 212–213) chides a friend for smashing some gourds he sees as worthless, for they are too big to serve as water containers or dippers. As he says, "Why not lash them together like big buoys and go floating on the rivers and lakes instead of worrying that they were too big to dip into anything?" Similarly, a gnarled old tree that is too crooked to serve as lumber can offer other kinds of value: "You have a big tree and are upset that you can't use it. Why not plant it by a nothing-at-all village in a wide empty waste? You could do nothing, dillydallying by its side, or nap, ho-hum, beneath it. It won't fall to any axe's chop and nothing will harm it. Since it isn't any use, what bad can happen to it?" (Zhuangzi 2001, p. 213). Zhuangzi's playful stories show us that new relationships—to gourds and trees, to one another, and to the natural world—are possible.

For further thought

1. What, in your view, would an ecological culture look like? What obstacles and opportunities do we face in enacting such a culture?
2. How might "bioregional cosmopolitanism" strike a balance between localism and globalism? Is this a good approach?
3. How might the perspectives of Confucianism and Daoism contribute to the development of an environmental ethic? Can these perspectives be useful complements to the Western tradition?

Conclusion

As noted in Chapter 1, environmental ethics is a fundamentally hopeful discipline. It asks us to reflect on our relationships with the natural world, and to work to improve those relationships. The processes and products of this reflection can take many forms, from formal philosophical argument, to thoughtful analysis of the values embedded in science, economics, and policy, to the redesign of individual and institutional practices. Thinking creatively and flexibly, from multiple perspectives, is key. Traditional philosophical modes of reflection, along with art, ritual, myth, and narrative, provide "technologies of the imagination" (Jordan et al. 2012) that have the potential to lay the groundwork for social change, and for a new environmental ethic. We need logical arguments, but not only logical arguments. There are changes ahead. Do not be too certain, do not hold too tight to what you think you know. Take things seriously, but not too seriously. As Gary Snyder (1990, p. 20) says, "We are all capable of extraordinary transformations."

Further reading

Bellah, R. N., Madsen, R., Sullivan, W. M., Swidler, A., and Tipton, S. M. (2008). *Habits of the Heart: Individualism and Commitment in American Life.* Berkeley, CA: University of California Press.

Carson, R. (1965). *The Sense of Wonder.* New York: Harper and Row.

Light, A. and Katz, E. (eds). (1996). *Environmental Pragmatism.* New York: Routledge.

Mulgan, T. (2011). *Ethics for a Broken World.* Montreal, Canada: McGill-Queen's University Press.
Norgaard, K. M. (2011). *Living in Denial: Climate Change, Emotions, and Everyday Life.* Cambridge, MA: MIT Press.
O'Neill, J., Holland, A., and Light, A. (2008). *Environmental Values.* New York: Routledge.
Saito, Y. (2007). *Everyday Aesthetics.* New York: Oxford University Press.
Snyder, G. (1990). *The Practice of the Wild.* San Francisco, CA: North Point Press.
Thomashow, M. (2002). *Bringing the Biosphere Home: Learning to Perceive Global Environmental Change.* Cambridge, MA: MIT Press.

Notes

Chapter 2

1 This example focuses on pleasures and pains, the concern of hedonistic utilitarianism. But preference utilitarianism works to commensurate values in the same basic way: all preferences (even ones we might find crazy or offensive) count.

2 The trolley problem is a thought experiment in which a train is hurtling down the track toward five people. You are standing in the proximity of a lever that can switch the train to an alternate track with one person on it. Should you pull the lever? We are to assume that the train will hit and kill whomever is on the track ahead of it.

Chapter 3

1 As J. Baird Callicott (1987, p. 197) puts it, "The contemporary animal liberation/rights, and [biocentric] ethics are, at bottom, simply direct applications of the modern classical paradigm of moral argument. But this standard modern model of ethical theory provides no possibility whatever for the moral consideration of wholes . . . since wholes per se have no psychological experience of any kind. Because mainstream moral theory has been 'psychocentric,' it has been radically and intractably individualistic or 'atomistic' in its fundamental theoretical orientation."

2 McShane (2007b, p. 53) suggests that awe is an intrinsically valuing attitude as well, but this seems more controversial than does reverence. One might be awed by the latest smartphone, but it is not clear that being awestruck by a particular technological device implies that one intrinsically values it.

3 On a related note, Nel Noddings (2003, pp. 96–97) argues that "caring is not in itself a virtue . . . We must not reify virtues and turn our caring toward them. If we do this, our ethic turns inward . . . The fulfillment of virtue is both in me and in the other."

Chapter 6

1 Defining "a minimally decent life" is, of course, tricky and controversial. Here, we can follow the lead of Henry Shue (2010, p. 108) who suggests that "having enough" should be understood as "enough for a decent chance for a reasonably healthy and active life of more or less normal length"—more than enough for bare physical survival, enough to live "a distinctively human, if modest, life."

References

Adams, C.J. (1990). *The Sexual Politics of Meat: A Feminist-Vegetarian Critical Theory*. New York: Continuum.

Adger, W. N. (2001). "Scales of governance and environmental justice for adaptation and mitigation of climate change." *Journal of International Development*, 13(7), 921–931.

Adger, W. N., Arnell, N. W., and Tompkins, E. L. (2005). "Successful adaptation to climate change across scales." *Global Environmental Change*, 15(2), 77–86.

Agyeman, J. (2013). *Introducing Just Sustainabilities: Policy, Planning, and Practice*. London: Zed Books.

Allen, C. (2011). "Animal consciousness," in E. N. Zalta (ed.), *The Stanford Encyclopedia of Philosophy*, Winter 2011 edition, http://plato.stanford.edu/archives/win2011/entries/consciousness-animal/.

American Planning Association. (2011). "Food policy councils: Food system planning briefing paper," http://www.planning.org/nationalcenters/health/briefingpapers/pdf/foodcouncils.pdf.

Andreou, C. (2007). "Environmental preservation and second-order procrastination." *Philosophy and Public Affairs*, 35(3), 233–248.

Aristotle. (1999). *Nicomachean Ethics* (2nd edn). Translated by Terence Irwin. Indianapolis, IN: Hackett.

Attfield, R. (2003). *Environmental Ethics: An Overview for the Twenty-First Century*. Cambridge, England: Polity Press.

Audi, R. and Murphy, P. E. (2006). "The many faces of integrity." *Business Ethics Quarterly*, 16(1), 3–21.

Baatz, C. (2014). "Climate change and individual duties to reduce GHG emissions." *Ethics, Policy & Environment*, 17(1), 1–19.

Baer, P. (with T. Athanasiou, S. Kartha, and E. Kemp-Benedict). (2010). "Greenhouse development rights: a framework for climate protection that is 'more fair' than equal per capita emissions rights," in S. Gardiner, S. Caney, D. Jamieson, and H. Shue (eds), *Climate Ethics: Essential Readings*. New York: Oxford, pp. 215–230.

Becker, C. U. (2012). *Sustainability Ethics and Sustainability Research*. New York: Springer.

Bellah, R. N., Madsen, R., Sullivan, W. M., Swidler, A., and Tipton, S. M. (2008). *Habits of the Heart: Individualism and Commitment in American Life*. Berkeley, CA: University of California Press.

Berry, W. (1997). *The Unsettling of America: Culture and Agriculture*. San Francisco, CA: Sierra Club Books.

Blatt, H. (2008). *America's Food: What You Don't Know about What You Eat*. Cambridge, MA: MIT Press.

Blum, L. (1991). "Moral perception and particularity." *Ethics,* 101, 701–725.
Bodeen, C. (2007). "In 'e-waste' heartland, a toxic China." *New York Times,* November 18, 2007, http://www.nytimes.com/2007/11/18/world/asia/18iht-waste.1.8374259.html.
Bookchin, M. (1995). *Re-enchanting Humanity: A Defense of the Human Spirit against Antihumanism, Misanthropy, Mysticism, and Primitivism.* New York: Cassell.
— (2002). "Social ecology versus deep ecology," in D. Schmidtz and E. Willott (eds), *Environmental Ethics: What Really Matters, What Really Works.* New York: Oxford University Press, pp. 126–136.
Booth, C. (2009). "A motivational turn for environmental ethics." *Ethics and the Environment,* 14(1), 53–78.
Borgmann, A. (1984). *Technology and the Character of Contemporary Life.* Chicago: University of Chicago Press.
— (2006). *Real American Ethics: Taking Responsibility for Our Country.* Chicago: University of Chicago Press.
Botkin, D. (1990). *Discordant Harmonies: A New Ecology for the Twenty-First Century.* New York: Oxford University Press.
— (1996). "Adjusting law to nature's discordant harmonies." *Duke Environmental Law and Policy Forum,* 7(25), 25–38.
Bradwell v. Illinois. 1872. 83 U.S. 130.
Brennan, A. (1992). "Moral pluralism and the environment." *Environmental Values,* 1(1), 15–33.
Bullard, R.D., Mohai, P., Saha, R., and Wright, B. (2008). "Toxic wastes and race at twenty: Why race still matters after all of these years." *Environmental Law,* 38, 371–411.
Cafaro, P. (2001). "Thoreau, Leopold, and Carson: Toward an environmental virtue ethics." *Environmental Ethics,* 22, 3–17.
— (2005). "Gluttony, arrogance, greed, and apathy: An exploration of environmental vice," in R. Sandler and P. Cafaro (eds), *Environmental Virtue Ethics.* Lanham, MD: Rowman & Littlefield, pp. 135–158.
Cahen, H. (2003). "Against the moral considerability of ecosystems," in A. Light and H. Rolston III (eds), *Environmental Ethics: An Anthology.* Malden, MA: Blackwell, pp. 114–128.
Callahan, D. (1971). "What obligations do we have to future generations?" *American Ecclesiastical Review,* 164(4), 265–280.
Callicott, J. B. (1983). "The land aesthetic." *Environmental Review,* 7(4), 345–358.
— (1987). "The conceptual foundations of the land ethic," in J. B. Callicott (ed.), *Companion to A Sand County Almanac: Interpretive and Critical Essays.* Madison, WI: University of Wisconsin Press, pp. 186–217.
— (1999a). "Rolston on intrinsic value: A deconstruction," in *Beyond the Land Ethic: More Essays in Environmental Philosophy.* Albany, NY: State University of New York Press, pp. 221–237.
— (1999b). "Holistic environmental ethics and the problem of ecofascism," in *Beyond the Land Ethic: More Essays in Environmental Philosophy.* Albany, NY: State University of New York Press, pp. 59–76.
Carlson, A. (2012). "Environmental aesthetics," in E. N. Zalta (ed.), *The Stanford Encyclopedia of Philosophy,* Summer 2012 edition, http://plato.stanford.edu/archives/sum2012/entries/environmental-aesthetics/.

Carroll, C. (2008). "High-tech trash." *National Geographic*, January 2008, http://ngm.nationalgeographic.com/2008/01/high-tech-trash/carroll-text.

Carson, R. (1965). *The Sense of Wonder*. New York: Harper & Row.

Carter, A. (2011). "Toward a multidimensional, environmentalist ethic." *Environmental Values*, 20, 347–374.

Carter, R. (2009). "Watsuji Tetsurô," in E. N. Zalta (ed.), *The Stanford Encyclopedia of Philosophy*, Spring 2013 edition, http://plato.stanford.edu/archives/spr2013/entries/watsuji-tetsuro/.

Chen, Y., Ebenstein, A., Greenstone, M., and Li, H. (2013). "Evidence on the impact of sustained exposure to air pollution on life expectancy from China's Huai River policy." *Proceedings of the National Academy of Sciences*, 110(32), 12936–12941.

Choi, Y. D. (2007). "Restoration ecology to the future: A call for new paradigm." *Restoration Ecology*, 15(2), 351–353.

Clewell, A. (2000). "Restoring for natural authenticity." *Ecological Restoration*, *18*(4), 216–217.

Commission for Racial Justice, United Church of Christ. (1987). "Toxic wastes and race in the United States," http://www.ucc.org/about-us/archives/pdfs/toxwrace87.pdf.

Cooper, G. J. (2003). *The Science of the Struggle for Existence: On the Foundations of Ecology*. New York: Cambridge University Press.

Corbett, J. B. (2001). "Women, scientists, agitators: Magazine portrayal of Rachel Carson and Theo Colburn." *Journal of Communication*, 51(4), 720–749.

Corner, A., and Pidgeon, N. (2010). "Geoengineering the climate: the social and ethical implications." *Environment: Science and Policy for Sustainable Development*, 52(1), 24–37.

Council on Environmental Quality (CEQ). (1997). "Environmental justice: Guidance under the National Environmental Policy Act." http://ceq.hss.doe.gov/nepa/regs/ej/justice.pdf.

Coveney, J. and O'Dwyer, L. A. (2009). "Effects of mobility and location on food access." *Health and Place*, 15(1), 45–55.

Cronon, W. (1995). "The trouble with wilderness; Or, getting back to the wrong nature," in W. Cronon (ed.), *Uncommon Ground: Toward Reinventing Nature*. New York: W.W. Norton, pp. 69–90.

Crutzen, P. J. and Schwägerl, C. (2011). "Living in the Anthropocene: Toward a new global ethos." *Yale Environment*, 360, January 24, 2011, http://e360.yale.edu/feature/living_in_the_anthropocene_toward_a_new_global_ethos/2363/.

Culotta, E. (2005). "Chimp genome catalogs differences with humans." *Science*, 309, 1468–1469.

Cuomo, C. J. (1998). *Feminism and Ecological Communities*. NY: Routledge.

Curtin, D. (2004). "Contextual moral vegetarianism," in S. F. Sapontzis (ed.), *Food for Thought: The Debate Over Eating Meat*. Amherst, NY: Prometheus Books, pp. 272–283.

Cypher, J. and Higgs, E. (1997). "Colonizing the imagination: Disney's wilderness lodge." *Capitalism Nature Socialism*, 8(4), 107–130, doi: 10.1080/10455759709358768.

Dancy, J. (2009). "Moral particularism," in E. N. Zalta (ed.), *The Stanford Encyclopedia of Philosophy*, Spring 2009 edition, http://plato.stanford.edu/archives/spr2009/entries/moral-particularism/.

Daniels, N. (2011). "Reflective equilibrium," in E. N. Zalta (ed.), *The Stanford Encyclopedia of Philosophy,* Spring 2011 edition, http://plato.stanford.edu/archives/spr2011/entries/reflective-equilibrium/.

Darwin, C. (2001). *On the Origin of Species*. Cambridge, MA: Harvard University Press.

Davion, V. (1998). "How feminist is ecofeminism?" in D. VanDeVeer and C. Pierce (eds), *The Environmental Ethics and Policy Book* (2nd edn). Belmont, CA: Wadsworth, pp. 278–285.

Dawe, N. K. and Ryan, K. L. (2003). "The faulty three-legged stool model of sustainable development." *Conservation Biology,* 17(5), 1458–1460.

De-Shalit, A. (1995). *Why Posterity Matters: Environmental Policies and Future Generations*. London: Routledge.

Denevan, W. M. (1992). "The pristine myth: The landscape of the Americas in 1492." *Annals of the Association of American Geographers,* 82(3), 369–385.

DesJardins, J. R. (2006). *Environmental Ethics: An Introduction to Environmental Philosophy* (4th edn). Belmont, CA: Wadsworth.

Diffenbaugh, N.S. and Field, C. B. (2013). "Changes in ecologically critical terrestrial climate conditions." *Science,* 341(6145), 486–492.

Dillon, R. S. (2010). "Respect," in E. N. Zalta (ed.), *The Stanford Encyclopedia of Philosophy,* Fall 2010 edition, http://plato.stanford.edu/archives/fall2010/entries/respect/.

Donlan, C. J., Berger, J., Bock, C. E., Bock, J. H., Burney, D. A., Estes, J. A., et al. (2006). "Pleistocene rewilding: An optimistic agenda for twenty-first century conservation." *American Naturalist,* 168(5), 660–681.

Driver, J. (2004). *Uneasy Virtue*. New York: Cambridge University Press.

Earth Charter Initiative. "The Earth charter," http://www.earthcharterinaction.org/content/pages/Read-the-Charter.html.

Egan, D. (2006). "Authentic ecological restoration." *Ecological Restoration,* 24(4), 223–224.

Ehrenfeld, D. (1988). "Why put a value on biodiversity?" in E. O. Wilson and F. M. Peter (eds), *Biodiversity.* Washington, DC: National Academy Press, pp. 212–216.

Elliot, R. (1982). "Faking nature." *Inquiry, 25,* 1–93.

Ellis, E. (2011). "The planet of no return." *Breakthrough Journal,* 2(Fall 2011), http://thebreakthrough.org/index.php/journal/past-issues/issue-2/the-planet-of-no-return.

Endres, D. (2012). "Sacred land or national sacrifice zone: The role of values in the Yucca Mountain participation process." *Environmental Communication: A Journal of Nature and Culture,* doi:10.1080/17524032.2012.688060.

Engel, M. (2000). "The immorality of eating meat," in L. Pojman (ed.), *The Moral Life.* New York: Oxford University Press, pp. 856–890.

Feinberg, J. (2013). "The rights of animals and unborn generations," in R. Shafer-Landau (ed.), *Ethical Theory: An Anthology.* Malden, MA: John Wiley & Sons, pp. 372–380.

Fesmire, S. (2003). *John Dewey and Moral Imagination*. Bloomington, IN: Indiana University Press.

Foot, P. (1978). "The problem of abortion and the doctrine of the double effect," in *Virtues and Vices and Other Essays in Moral Philosophy.* Berkeley, CA: University of California Press, pp. 19–32.

Gamborg, C., Gremmen, B., Christiansen, S. B., and Sandoe, P. (2010). "De-domestication: Ethics at the intersection of landscape restoration and animal welfare." *Environmental Values,* 19(1), 57–78.

Gardiner, S. (2006). "A perfect moral storm: Climate change, intergenerational ethics and the problem of moral corruption." *Environmental Values,* 15(3), 397–413.

Gillis, J. (2013). "Heat-trapping gas passes milestone, raising fears." *New York Times,* May 10, http://www.nytimes.com/2013/05/11/science/earth/carbon-dioxide-level-passes-long-feared-milestone.html?pagewanted=all.

Glazebrook, T. (2002). "Karen Warren's ecofeminism." *Ethics and the Environment,* 7, 12–26.

Goldstein, B. E. and Butler, W. H. (2010). "The US fire learning network: Providing a narrative framework for restoring ecosystems, professions, and institutions." *Society and Natural Resources,* 23(10), 935–951.

Goodin, R. (1998). "Communities of enlightenment." *British Journal of Political Science,* 28(3), 531–558.

Goodpaster, K. (1978). "On being morally considerable." *Journal of Philosophy,* 75(6), 308–325.

Gowans, C. (2012). "Moral relativism," in E. N. Zalta (ed.), *The Stanford Encyclopedia of Philosophy,* Spring 2012 edition, http://plato.stanford.edu/archives/spr2012/entries/moral-relativism/.

Gross, M. (2010). *Ignorance and Surprise: Science, Society, and Ecological Design.* Cambridge, MA: MIT Press.

Gruen, L. (1993). "Dismantling oppression: An analysis of the connection between women and animals," in G. Gaard (ed.), *Ecofeminism: Women, Animals, and Nature.* Philadelphia: Temple University Press, pp. 60–90.

Gruen, L., Jamieson, D., and Schlottman, C. (2013). *Reflecting on Nature: Readings in Environmental Ethics and Philosophy.* New York: Oxford University Press.

Guha, R. (1989). "Radical American environmentalism and wilderness preservation: A third world critique." *Environmental Ethics,* 11, 71–83.

Halfon, M. (1989). *Integrity: A Philosophical Inquiry.* Philadelphia: Temple University.

Hall, M. (2005). *Earth Repair: A Transatlantic History of Environmental Restoration.* Charlottesville, VA: University of Virginia Press.

Hardin, G. (1968). "The tragedy of the commons." *Science,* 162, 1243–1248.

Harding, S. (1991). *Whose Science? Whose Knowledge? Thinking from Women's Lives.* Ithaca, NY: Cornell University Press.

Harris, P. G. (2008). "Climate change and global citizenship." *Law and Policy,* 30(4), 481–501.

Havlick, D. G., Hourdequin, M., and John, M. (2014). "Examining restoration goals at a former military site: the Rocky Mountain Arsenal, Colorado (USA)." *Nature and Culture,* 9(3).

Higgs, E. (2003). *Nature by Design: People, Natural Process, and Ecological Restoration.* Cambridge, MA: MIT Press.

Hill Jr., T. E. (1983). "Ideals of human excellence and preserving natural environments." *Environmental Ethics,* 5, 211–224.

— (2012). *Virtue, Rules, and Justice: Kantian Aspirations.* New York: Oxford University Press.

Hobbs, R. J., Hallett, L. M., Ehrlich, P. R., and Mooney, H. A. (2011). "Intervention ecology: Applying ecological science in the twenty-first century." *BioScience* 61(6): 442–450.

Hobbes, Thomas. (1968). *Leviathan.* New York: Penguin.

Holland, A. (2012). "The value space of meaningful relations," in E. Brady and P. Phemister (eds), *Human–Environment Relations: Transformative Values in Theory and Practice.* New York: Springer, pp. 3–15.

Hourdequin, M. (2010). "Climate, collective action and individual ethical obligations." *Environmental Values,* 19(4), 443–464.

— (2012). "Climate, solidarity, and moral risk," in C. J. Preston (ed.), *Engineering the Climate: The Ethics of Solar Radiation Management.* Lanham, MD: Lexington Books, pp. 15–32.

— (2013). "Restoration and history in a changing world: A case study in ethics for the Anthropocene." *Ethics and the Environment,* 18(2), 115–134.

Hourdequin, M. and Havlick, D. G. (2013). "Restoration and authenticity revisited." *Environmental Ethics,* 35(1), 79–93.

Hourdequin, M. and Wong, D. (2005). "A relational approach to environmental ethics." *Journal of Chinese Philosophy,* 32(1), 19–33.

Hulme, M. (2009). *Why We Disagree about Climate Change: Understanding Controversy, Inaction and Opportunity.* New York: Cambridge University Press.

Hursthouse, R. (2001). *On Virtue Ethics.* New York: Oxford University Press.

— (2012). "Virtue ethics," in E. N. Zalta (ed.), *The Stanford Encyclopedia of Philosophy,* Summer 2012 edition, http://plato.stanford.edu/archives/sum2012/entries/ethics-virtue/.

Intergovernmental Panel on Climate Change (IPCC). (2007a). "FAQ 6.2: Is the current rate of climate change unusual compared to other changes in the Earth's history?" *Climate Change 2007: Working Group I: The Physical Science Basis.* https://www.ipcc.ch/publications_and_data/ar4/wg1/en/faq-6–2.html.

— (2007b). *Climate Change 2007: Synthesis Report. Contribution of Working Groups I, II, and II to the Fourth Assessment Report of the Intergovernmental Panel on Climate Change,* core writing team, R. K. Pachauri and A. Reisinger (eds). Geneva, Switzerland: IPCC.

— (2007c). "Summary for policymakers." *Climate Change 2007: Impacts, Adaptation and Vulnerability. Contribution of Working Group II to the Fourth Assessment Report of the Intergovernmental Panel on Climate Change,* edited by M. L. Parry, O. F. Canziani, J. P. Palutikof, P. J. van der Linden, and C. E. Hanson. Cambridge, UK: Cambridge University Press, pp. 7–22.

— (2014). *Climate Change 2014: Impacts, Adaptation and Vulnerability, Vol. II: Regional Impacts,* https://www.ipcc.ch/report/ar5/wg2/.

Ivanhoe, P. J. and Van Norden, B. W. (2001). *Readings in Classical Chinese Philosophy,* second edition. Indianapolis, IN: Hackett.

Jackson, W. (1994). *Becoming Native to this Place.* Lexington, KY: University Press of Kentucky.

Jacobs, M. (1999). "Sustainable development as a contested concept," in A. Dobson (ed.), *Fairness and Futurity: Essays on Environmental Sustainability and Social Justice.* New York: Oxford University Press, pp. 21–45.

Jacobson, D. (2005). "Seeing by feeling: Virtues, skills, and moral perception." *Ethical Theory and Moral Practice*, 8, 387–409.

Jamieson, D. (1996). "Ethics and intentional climate change." *Climatic Change*, 33(3), 323–336.

— (2007). "The moral and political challenges of climate change," in S. C. Moser and L. Dilling (eds), *Creating a Climate for Change: Communicating Climate Change and Facilitating Social Change*. New York: Cambridge University Press, pp. 475–482.

— (2010). "Climate change, responsibility, and justice." *Science and Engineering Ethics*, 16, 441–445.

Johnson, B. (2003). "Ethical obligations in a tragedy of the commons." *Environmental Values*, *12*(3), 271–287.

Jordan III, W. R. (2003). *The Sunflower Forest: Ecological Restoration and the New Communion with Nature*. Berkeley, CA: University of California Press.

Jordan III, W. R., Barrett, N. F., Curtis, K., Heneghan, L., Honold, R., and LeVasseur, T. (2012). "Foundations of conduct." *Environmental Ethics*, 34(3), 291–312.

Jordan III, W. R. and Lubick, G. M. (2011). *Making Nature Whole: A History of Ecological Restoration*. Washington, DC: Island Press.

Kahn, P. H. (2011). *Technological Nature: Adaptation and the Future of Human Life*. Cambridge, MA: MIT Press.

Kant, I. (1995). *Foundations of the Metaphysics of Morals*. Edited by L. W. Beck. Upper Saddle River, NJ: Prentice-Hall.

Karl, T. R., Meehl, G. A., Peterson, T. C., Kunkel, K. E., Gutowski, Jr., W. J., and Easterling, D. R. (2008). "Executive summary," in T. R. Karl, G. A. Meehl, C. D. Miller, S. J. Hassol, A. M. Waple, and W. L. Murray (eds), *Weather and Climate Extremes in a Changing Climate. Regions of Focus: North America, Hawaii, Caribbean, and U.S. Pacific Islands*. Washington, DC: U.S. Climate Change Science Program and the Subcommittee on Global Change Research, pp. 1–9.

Kasulis, T. P. (2002). *Intimacy or Integrity: Philosophy and Cultural Difference*. Honolulu, HI: University of Hawaii Press.

Kates, R. W., Parris, T. M., and Leiserowitz, A. A. (2005). "What is sustainable development? Goals, indicators, values, and practice." *Environment: Science and Policy for Sustainable Development*, 47(3), 8–21.

Katz, E. (1997). "The big lie: Human restoration of nature," in *Nature as Subject: Human Obligation and Natural Community*. New York: Rowman & Littlefield, pp. 93–108.

Keane, R. E., Hessburg, P. F., Landres, P. B., and Swanson, F. J. (2009). "The use of historical range and variability (HRV) in landscape management." *Forest Ecology and Management*, 258(7), 1025–1037.

Kheel, M. (2008). *Nature Ethics: An Ecofeminist Perspective*. Lanham, MD: Rowman & Littlefield.

Killingsworth, M. J. and Palmer, J. S. (1995). "The discourse of 'environmentalist hysteria.'" *Quarterly Journal of Speech*, 81(1), 1–19.

King, R. J. H. (2000). "Environmental ethics and the built environment." *Environmental Ethics*, *22*, 115–131.

Kitcher, P. (2001). *Science, Truth, and Democracy*. New York: Oxford University Press.

Knights, P. (2012). *Consumption, Environment and Ethics: An Analysis of Moral and Welfare Arguments for Reducing Personal Consumption*. Ph.D. thesis, Department of Philosophy, University of Manchester, United Kingdom.

Kolbert, E. (2012). "Recall of the wild: A quest to engineer a world before humans." *New Yorker Magazine*, December 24, pp. 50–60.

Krauss, C. (1994). "Women of color on the front line," in R. Bullard (ed.), *Unequal Protection: Environmental Justice and Communities of Color*. San Francisco, CA: Sierra Club Books, pp. 256–271.

Kuebbeler, M., Lohmann, U., and Feichter, J. (2012). "Effects of stratospheric sulfate aerosol geo-engineering on cirrus clouds." *Geophysical Research Letters*, 39, L23803, doi:10.1029/2012GL053797

Kuhn, T. (1996). *The Structure of Scientific Revolutions*. Chicago: University of Chicago Press.

Lai, K. (2006). *Learning from Chinese Philosophies: Ethics of Interdependent and Contextualised Self*. Burlington, VT: Ashgate.

Larsen, B. S. and Tufte, T. (2003). "Rituals in the modern world: Applying the concept of ritual in media ethnography," in P. D. Murphy and M. M. Krady (eds), *Global Media Studies—Ethnographic Perspectives*. New York: Routledge, pp. 90–106.

Lau, D. C. (trans.). (1979). *Confucius: The Analects*. London: Penguin.

Lear, G. R. (2013). "Aristotle," in H. LaFollette (ed.), *The International Encyclopedia of Ethics*. New York: Blackwell, pp. 348–362.

LeGuin, U. (2004). *The Wind's Twelve Quarters: Stories* (originally published in 1975). New York: HarperCollins.

Leopold, A. (1949). *A Sand County Almanac and Sketches Here and There*. New York: Oxford University Press.

Leopold, A. C. (2004). "Living with the land ethic." *BioScience*, 54(2), 149–154.

Leyden, K. M. (2003). "Social capital and the built environment: The importance of walkable neighborhoods." *American Journal of Public Health*, 93(9), 1546–1551.

Light, A. (2002). "Contemporary environmental ethics from metaethics to public philosophy." *Metaphilosophy*, 33(4), 426–449.

— (2003). "Ecological restoration and the culture of nature: A pragmatic perspective," in A. Light and H. Rolston III (eds), *Environmental Ethics: An Anthology*. Malden, MA: Blackwell, pp. 398–411.

— (2006). "Ecological citizenship: The democratic promise of restoration," in R. H. Platt (ed.), *The Humane Metropolis: People and Nature in the 21st-Century City*. Amherst, MA: University of Massachusetts Press, pp. 169–181.

Light, A. and Katz, E. (1996). "Introduction: Environmental pragmatism and environmental ethics as contested terrain," in A. Light and E. Katz (eds), *Environmental Pragmatism*. New York: Routledge, pp. 1–18.

Lomborg, B. (2001). *The Skeptical Environmentalist: Measuring the Real State of the World*. New York: Cambridge University Press.

Longino, H. (1990). *Science as Social Knowledge*. Princeton, NJ: Princeton University Press.

Louv, R. (2005). *Last Child in the Woods: Saving Our Children from Nature-Deficit Disorder*. Chapel Hill, NC: Algonquin Books.

— (2011). *The Nature Principle: Human Restoration and the End of Nature-Deficit Disorder*. Chapel Hill, NC: Algonquin Books.

Mallory, C. (2006). "Ecofeminism and forest defense in Cascadia: Gender, theory and radical activism." *Capitalism Nature Socialism*, 17, 32–49.
McDonald, N. C., Brown, A. L., Marchetti, L. M., and Pedroso, M. S. (2011). "US school travel, 2009: An assessment of trends." *American Journal of Preventive Medicine*, 41(2), 146–151.
McDowell, J. (1998). *Mind, Value, and Reality*. Cambridge, MA: Harvard University Press.
McKibben, B. (1989). *The End of Nature*. New York: Random House.
McShane, K. (2007a). "Rolston's theory of value," in C. J. Preston and W. Ouderkirk (eds), *Nature, Value, Duty: Life on Earth with Holmes Rolston, III*. Dordrecht, The Netherlands: Springer, pp. 1–15.
— (2007b). "Why environmental ethics shouldn't give up on intrinsic value." *Environmental Ethics*, 29, 43–61.
— (2012). "Some challenges for narrative accounts of value." *Ethics and the Environment, 17*, 45–69.
Merchant, C. (1989). *The Death of Nature: Women, Ecology, and the Scientific Revolution*. New York: Harper & Row.
— (1992). *Radical Ecology: The Search for a Livable World*. New York: Routledge.
Meyer, L. (2010). "Intergenerational justice," in E. N. Zalta (ed.), *The Stanford Encyclopedia of Philosophy*, Spring 2010 edition, http://plato.stanford.edu/archives/spr2010/entries/justice-intergenerational/.
Midgley, M. (1996). "Sustainability and moral pluralism." *Ethics and the Environment*, 1(1), 41–54.
Mies, M. and Shiva, V. (1993). *Ecofeminism*. New York: Zed Books.
Millennium Ecosystem Assessment. (2003). *Ecosystems and Human Well-Being: A Framework for Assessment*. Washington, DC: Island Press.
Minteer, B. A. (1998). "No experience necessary? Foundationalism and the retreat from culture in environmental ethics." *Environmental Values*, 7(3), 333–348.
Minteer, B. A. and Manning, R. E. (2000). "Convergence in environmental values: An empirical and conceptual defense." *Ethics, Place, and Environment*, 3(1), 47–60.
Muir, J. (2001). *The Wilderness World of John Muir*, edited by Edwin Way Teale. New York: Houghton Mifflin Harcourt.
Mulgan, T. (2011). *Ethics for a Broken World*. Montreal, Canada: McGill-Queen's University Press.
Naess, A. (2005). "Self-realization: An ecological approach to being in the world," in A. R. Dregson (ed.), *The Selected Works of Arne Naess*. Dordrecht, The Netherlands: Springer, pp. 515–530.
— (2012). "Ecosophy T: Deep versus shallow ecology," in L. P. Pojman and P. Pojman (eds), *Environmental Ethics: Readings in Theory and Practice* (6th edn). Boston, MA: Wadsworth, pp. 133–142.
Naess, A. and Sessions, G. (1995). "Platform principles of the deep ecology movement," in A. R. Dregson and U. Inoue (eds), *The Deep Ecology Movement: An Introductory Anthology*. Berkeley, CA: North Atlantic Books, pp. 49–53.

Napier, M. A., Brown, B. B., Werner, C. M., and Gallimore, J. (2011). "Walking to school: Community design and child and parent barriers." *Journal of Environmental Psychology,* 31(1), 45–51.
Nash, R. F. (2014). *Wilderness and the American Mind.* New Haven: Yale University Press.
Nathanson, S. (2012). "Ethics for a broken world: Imagining philosophy after catastrophe." *Notre Dame Philosophical Reviews*, September 10, 2012, http://ndpr.nd.edu/news/33196-ethics-for-a-broken-world-imagining-philosophy-after-catastrophe/.
Newton, J. L. and Freyfogle, E. T. (2005). "Sustainability: A dissent." *Conservation Biology,* 19(1), 23–32.
NOAAa. "What is ocean acidification?" National Oceanic and Atmospheric Administration, PMEL Carbon Program. http://www.pmel.noaa.gov/co2/story/What+is+Ocean+Acidification%3F.
NOAAb. "Human and economic indicators—Shishmaref." Arctic Change: A Near-Real Time Arctic Change Indicator Website. http://www.arctic.noaa.gov/detect/human-shishmaref.shtml.
Noddings, N. (2003). *Caring: A Feminist Approach to Ethics and Moral Education.* Berkeley, CA: University of California Press.
Nolt, J. (2011). "Nonanthropocentric climate ethics." *WIREs Climate Change*, 2, 701–711.
Nordhaus, W. (2007). "Critical assumptions in the Stern Review on climate change." *Science, 317*(5835), 201–202.
— (2008). *A Question of Balance: Weighing the Options on Global Warming Policies.* New Haven, CT: Yale University Press.
Norgaard, K. M. (2011). *Living in Denial: Climate Change, Emotions, and Everyday Life.* Cambridge, MA: MIT Press.
Norton, B. (1991). *Toward Unity Among Environmentalists.* New York: Oxford University Press.
Norton, B. G. (1996). "Integration or reduction: Two approaches to environmental value," in A. Light and E. Katz (eds), *Environmental Pragmatism.* New York: Routledge, pp. 105–130.
— (2005). *Sustainability: A Philosophy of Adaptive Ecosystem Management.* Chicago: University of Chicago Press.
Nozick, R. (1974). *Anarchy, State, and Utopia.* New York: Basic Books.
Nussbaum, M. (2000). *Women and Human Development: The Capabilities Approach.* New York: Cambridge University Press.
Oliver, M. (1997). "Have you ever tried to enter the long black branches," in *West Wind.* New York: Houghton Mifflin, pp. 61–63.
O'Neill, J. (1993). "Future generations: Present harms." *Philosophy,* 68, 35–51.
O'Neill, J., Holland, A., and Light, A. (2008). *Environmental Values.* New York: Routledge.
Ostrom, E. (2010). "Polycentric systems for coping with collective action and global environmental change," *Global Environmental Change,* 20(4), 550–557.
Page, E. A. (2007). "Intergenerational justice of what: Welfare, resources or capabilities?" *Environmental Politics,* 16(3), 453–469.
Palmer, C. (2010). *Animal Ethics in Context.* New York: Columbia University Press.
Parfit, D. (1984). *Reasons and Persons.* New York: Oxford University Press.

Parker, K. A. (1996). "Pragmatism and environmental thought," in A. Light and E. Katz (eds), *Environmental Pragmatism*. New York: Routledge, pp. 21–37.
Partridge, E. (2001). "Future generations," in D. Jamieson (ed.), *A Companion to Environmental Philosophy*. Malden, MA: Blackwell, pp. 377–389.
Pellow, D. N. (2007). *Resisting Global Toxics: Transnational Movements for Environmental Justice*. Cambridge, MA: MIT Press.
Pethick, J. S. and Crooks, S. (2000). "Development of a coastal vulnerability index: A geomorphological perspective." *Environmental Conservation*, 27(4), 359–367.
Plumwood, V. (1996). "Nature, self, and gender: Feminism, environmental philosophy, and the critique of rationalism," in K. J. Warren (ed.), *Ecological Feminist Philosophies*. Bloomington, IN: Indiana University Press, pp. 155–180.
Preston, C. J. (ed.). (2012). *Engineering the Climate: The Ethics of Solar Radiation Management*. Lanham, MD: Lexington Books.
Purdy, J. (2013). "Our place in the world: A new relationship for environmental ethics and law." *Duke Law Journal*, 62(4), 857–932.
Quintero, I. and Wiens, J. J. (2013). "Rates of projected climate change dramatically exceed past rates of climatic niche evolution among vertebrate species." *Ecology Letters*, 16(8), 1095–1103.
Radick, G. (2003). "Is the theory of natural selection independent of its history?" in J. Hodge and G. Radick (eds), *The Cambridge Companion to Darwin*. New York: Cambridge University Press, pp. 143–167.
Rawls, J. (1971). *A Theory of Justice*. Cambridge, MA: Harvard University Press.
Rayner, S., Heyward, C., Kruger, T., Pidgeon, N., Redgwell, C., and Savulescu, J. (2013). "The Oxford Principles." *Climatic Change*, 121(3), 499–512.
Regan, T. (2004). *The Case for Animal Rights*. Berkeley, CA: University of California Press.
— (2012). "The radical egalitarian case for animal rights," in L. P. Pojman and P. Pojman (eds), *Environmental Ethics: Readings in Theory and Practice* (6th edn). Boston, MA: Wadsworth, pp. 81–88.
Riedel, J., Schumann, K., Kaminski, J., Call, J., and Tomasello, M. (2008). "The early ontogeny of human–dog communication." *Animal Behaviour*, 75(3), 1003–1014.
Roach, C. (1996). "Loving your mother: On the woman–nature relation," in K. J. Warren (ed.), *Ecological Feminist Philosophies*. Bloomington, IN: Indiana University Press, pp. 52–65.
Robinson, J. (2004). "Squaring the circle? Some thoughts on the idea of sustainable development." *Ecological Economics*, 48(4), 369–384.
Rollin, B. E. (1995). *The Frankenstein Syndrome: Ethical and Social Issues in the Genetic Engineering of Animals*. New York: Cambridge University Press.
Rolston III, H. (1988). *Environmental Ethics: Duties to and Values in the Natural World*. Philadelphia, PA: Temple University Press.
— (1996). "Feeding people versus saving nature," in W. Aiken and H. LaFollette (eds), *World Hunger and Morality* (2nd edn). Englewood Cliffs, NJ: Prentice-Hall, pp. 248–267.
— (2003). "Value in nature and the nature of value," in A. Light and H. Rolston III (eds), *Environmental Ethics: An Anthology*. Malden, MA: Blackwell, pp. 143–153.

— (2005). "Environmental virtue ethics: Half the truth but dangerous as a whole," in R. Sandler and P. Cafaro (eds), *Environmental Virtue Ethics*. Lanham, MD: Rowman & Littlefield, pp. 61–78.
— (2012). *A New Environmental Ethics: The Next Millennium for Life on Earth*. New York: Routledge.
Rosemont, H. (1991). "Rights-bearing individuals and role-bearing persons," in M. I. Bockover (ed.), *Rules, Rituals, and Responsibility: Essays Dedicated to Herbert Fingarette*. LaSalle, IL: Open Court Press, pp. 71–102.
Rosenberg, A. and McShea, D. W. (2008). *Philosophy of Biology: A Contemporary Introduction*. New York: Routledge.
Routley, R. (2013). "Is there a need for a new, an environmental ethic?" in L. Gruen, D. Jamieson, and C. Schlottman (eds), *Reflecting on Nature: Readings in Environmental Ethics and Philosophy* (2nd edn). New York: Oxford University Press, pp. 41–46.
Royal Society, The. (2009). *Geoengineering the Climate: Science, Governance, and Uncertainty*. London: The Royal Society.
Sagoff, M. (2012). "At the shrine of our lady of Fatima, or why political questions are not all economic," in L. P. Pojman and P. Pojman (eds), *Environmental Ethics: Readings in Theory and Practice* (6th edn). Boston, MA: Wadsworth, pp. 669–678.
Saito, Y. (2007). *Everyday Aesthetics*. New York: Oxford University Press.
Samuelsson, L. (2010). "Environmental pragmatism and environmental philosophy." *Environmental Ethics*, 32(4), 405–415.
Sandler, R. (2005). "Introduction: Environmental virtue ethics," in R. Sandler and P. Cafaro (eds), *Environmental Virtue Ethics*. Lanham, MD: Rowman & Littlefield, pp. 1–12.
Sapolsky, R. (1990). "Stress in the wild." *Scientific American*, 262(1), 116–123.
Sarewitz, D. (2004). "How science makes environmental controversies worse." *Environmental Science and Policy*, 7, 385–403.
Scheffer, V. B. (1993). "The Olympic goat controversy: A perspective." *Conservation Biology*, 7(4), 916–919.
Schlosberg, D. (2007). *Defining Environmental Justice*. New York: Oxford University Press.
Schor, J. (2010). *Plenitude: The New Economics of True Wealth*. New York: Penguin Press.
Science and Environmental Health Network. (1998). "The Wingspread Consensus Statement on the precautionary principle," http://www.sehn.org/wing.html.
Shiva, V. (1989). *Staying Alive: Women, Ecology and Development*. Atlantic Highlands, NJ: Zed Books.
— (1998). "Women, ecology, and development," in D. VanDeVeer and C. Pierce (eds), *The Environmental Ethics and Policy Book* (2nd edn). Belmont, CA: Wadsworth, pp. 271–277.
— (2000). *Stolen Harvest*. Cambridge, MA: South End Press.
Shrader-Frechette, K. (2002). *Environmental Justice: Creating Equality, Reclaiming Democracy*. New York: Oxford University Press.
Shrader-Frechette, K. S. and McCoy, E. D. (1993). *Method in Ecology: Strategies for Conservation*. New York: Cambridge University Press.
Shue, H. (2001). "Climate," in D. Jamieson (ed.), *A Companion to Environmental Philosophy*. Malden, MA: Blackwell, pp. 449–459.

— (2010). "Global environment and international inequality," in S. M. Gardiner, S. Caney, D. Jamieson, and H. Shue (eds), *Climate Ethics: Essential Readings*. New York: Oxford University Press, pp. 101–111.

Singer, P. (2002). *Animal Liberation*. New York: HarperCollins.

— (2008). "A utilitarian defense of animal liberation," in L. P. Pojman and P. Pojman (eds), *Environmental Ethics: Readings in Theory and Application* (5th edn). Belmont, CA: Wadsworth, pp. 73–82

Singer, P. and Mason, J. (2006). *The Way We Eat: Why our Food Choices Matter.* Rodale Press.

Sinnott-Armstrong, W. (2005). "It's not *my* fault: Global warming and individual moral obligations," in W. Sinnott-Armstrong and R. Howarth (eds), *Perspectives on Climate Change: Science, Economics, Politics, Ethics*. Amsterdam: Elsevier, pp. 285–307.

Smith, M. B. (2001). "'Silence, Miss Carson!' Science, Gender, and the Reception of *Silent Spring*." *Feminist Studies*, 27(3), 733–752.

Snyder, G. (1990). *The Practice of the Wild*. San Francisco, CA: North Point Press.

Staatsbosbeheer (Dutch Forestry Commission). "Oostvaardersplassen: About the site." http://www.staatsbosbeheer.nl/English/Oostvaardersplassen/About%20the%20site.aspx

Stern, N. (ed.). (2007). *The Economics of Climate Change: The Stern Review*. Cambridge, UK: Cambridge University Press.

Stern, N. and Taylor, C. (2007). "Climate change: Risk, ethics, and the Stern Review." *Science*, 317(5835), 203–204.

Stone, C. D. (2010). "Should trees have standing? Toward legal rights for natural objects," in *Should Trees Have Standing? Law, Morality, and the Environment*. New York: Oxford University Press, pp. 1–31.

Taliaferro, C. (2001). "Early modern philosophy," in D. Jamieson (ed.), *A Companion to Environmental Philosophy*. Malden, MA: Blackwell, pp. 130–145.

Taylor, B. L. and Gerrodette, T. (1993). "The uses of statistical power in conservation biology: The Vaquita and northern Spotted Owl." *Conservation Biology*, 7, 489–500.

Taylor, P. (1981). "The ethics of respect for nature." *Environmental Ethics*, 3, 197–218.

Thomashow, M. (1999). "Toward a cosmopolitan bioregionalism," in M. V. McGinnis (ed.), *Bioregionalism*. New York: Routledge, pp. 121–132.

— (2002). *Bringing the Biosphere Home: Learning to Perceive Global Environmental Change*. Cambridge, MA: MIT Press.

Thomson, J. J. (1985). "The trolley problem." *Yale Law Journal*, 94, 1395–1415.

Throop, W. (2010). "Strengthening the social sustainability virtues: The role of philosophy." Paper presented at the International Society for Environmental Ethics Annual Meeting, Allenspark, Colorado.

Treanor, B. (2010). "Environmentalism and public virtue." *Journal of Agricultural and Environmental Ethics*, 23, 9–28.

Tuana, N. (2013). "Gendering climate knowledge for justice: Catalyzing a new research agenda," in M. Alston and K. Whittenbury (eds), *Research, Action and Policy: Addressing the Gendered Impacts of Climate Change*. Dordrecht: Springer, pp. 17–31.

Turkle, S. (2011). *Alone Together: Why We Expect More from Technology and Less from Each Other.* New York: Basic Books.

Van Der Heijden, H.-A. (2005). "Ecological restoration, environmentalism and the Dutch politics of 'new nature.'" *Environmental Values,* 14(4), 427–446.

Vucetich, J. A. and Nelson, M. P. (2010). "Sustainability: Virtuous or vulgar?" *BioScience,* 60(7), 539–544.

Waas, T., Hugé, J., Verbruggen, A., and Wright, T. (2011). "Sustainable development: A bird's eye view." *Sustainability,* 3(10), 1637–1661.

Walzer, M. (1987). *Interpretation and Social Criticism.* Cambridge, MA: Harvard University Press.

Warren, K. (1996). "The power and promise of ecological feminism," in K. J. Warren (ed.), *Ecological Feminist Philosophies.* Bloomington, IN: Indiana University Press, pp. 19–41.

Wenz, P. (2012). "Just garbage: The problem of environmental racism," in L. P. Pojman and P. Pojman (eds), *Environmental Ethics: Readings in Theory and Practice* (6th edn). Boston, MA: Wadsworth, pp. 530–539.

Weston, A. (1992). "Before environmental ethics." *Environmental Ethics,* 14, 321–338.

— (1996). "Beyond intrinsic value: Pragmatism in environmental ethics," in A. Light and E. Katz (eds), *Environmental Pragmatism.* New York: Routledge, pp. 285–306.

— (2009). *The Incompleat Eco-Philosopher.* Albany, NY: State University of New York Press.

— (2012). *Mobilizing the Green Imagination: An Exuberant Manifesto.* Gabriola Island, BC, Canada: New Society Publishers.

Whatmore, S. (2002). *Hybrid Geographies: Natures, Cultures, Spaces.* London: Sage.

White, R. (2004). "From wilderness to hybrid landscapes: The cultural turn in environmental history." *Historian,* 66(3), 557–564.

Whyte, K. P. (2012). "Indigenous peoples, solar radiation management, and consent," in C. J. Preston (ed), *Engineering the Climate: The Ethics of Solar Radiation Management.* Lanham, MD: Lexington, pp. 65–76.

Wiener, J. (2002). "Precaution in a multirisk world," in D. J. Paustenbach (ed.), *Human and Ecological Risk Assessment: Theory and Practice.* New York: John Wiley & Sons, pp. 1509–1531.

Wilson, E. J., Marshall, J., Wilson, R., and Krizek, K. J. (2010). "By foot, bus or car: Children's school travel and school choice policy." *Environment and Planning A,* 42(9), 2168–2185.

Wilson, E. O. (1984). *Biophilia.* Cambridge, MA: Harvard University Press.

Wilson, H. L. (2012). "The green Kant: Kant's treatment of animals," in L. P. Pojman and P. Pojman (eds), *Environmental Ethics: Readings in Theory and Practice* (6th edn). Boston, MA: Wadsworth, pp. 62–70.

Winne, M. (2009). "Fresh from . . . the city." *YES! Magazine,* February 13, http://www.yesmagazine.org/issues/food-for-everyone/fresh-from-. . .-the-city.

Wong, D. B. (2006). *Natural Moralities: A Defense of Pluralistic Relativism.* New York: Oxford University Press.

Wood, A. W. (1998). "Kant on duties regarding nonrational nature." *Proceedings of the Aristotelian Society,* Suppl. vols. 72, 189–228.

World Commission on Environment and Development (WCED). (1987). "Our common future." United Nations, http://www.un-documents.net/our-common-future.pdf.

Zhuangzi. (2001). "Zhuangzi," in P. J. Ivanhoe and B. W. Van Norden (eds), *Readings in Classical Chinese Philosophy* (2nd edn). Indianapolis, IN: Hackett, pp. 207–253.

Index